71 LESSONS FROM THE SKY

CIVILIAN HELICOPTERS

FLETCHER MCKENZIE

SSP

71 LESSONS FROM THE SKY

NEAR MISSES AND STORIES FROM
71 CIVILIAN HELICOPTER PILOTS
FROM AROUND THE WORLD

This edition published 2023 by Squabbling Sparrows Press

ISBN 978-0-4734930-80 (Paperback edition)
ISBN 978-0-9951170-44 (Ebook edition)

A catalogue record for this book is available from the National Library of New Zealand.

Published by Squabbling Sparrows Press
PO Box 4213, Marewa, Napier 4143
New Zealand

ALSO BY FLETCHER MCKENZIE

51 Lessons From The Sky (U.S. Air Force)

61 Lessons From The Sky (Military Helicopters)

71 Lessons From The Sky (Civilian Helicopters)

72 Lessons From The Sky (Cessna 172)

81 Lessons From The Sky (General Aviation)

101 Lessons From The Sky (Commercial Aviation)

TOPGUN Lessons From The Sky (U.S. Navy)

From The Pilot's Seat

Dedicated to three helicopter pioneers who helped shape the New Zealand helicopter industry.

Thank you
Sir Tim Wallis, Keith McKenzie and Bill Reid.

"The helicopter approaches closer than any other (vehicle) to fulfilment of mankind's ancient dream of the flying horse and the magic carpet."

Igor Sikorsky

CONTENTS

FOREWORD
CLAUDE VUICHARD

"Accidents are always the same; all that changes are the dates and registrations. The problem is between the headset."
Claude Vuichard

As a child I always dreamed to fly and I am lucky to have realised this dream. However there have been a number of times I also dreamed of being on the ground with my crew, this was during a number of rescues missions in the Swiss Alps when we were pushing the limits of the aircraft, flying in over 100 knot winds.

After forty years of flying, with 16,000 hours, I am still convinced I made the right choice. Aviation has helped me save lives, probably enough to fill a church, and that knowledge is good for my soul.

I love flying because of the third dimension and even though you may fly the same route many times, it never looks the same. My office always changes; it's probably the best desk you can have in the world.

I had a few close calls flying in the mountains, again when we were flying at the power limits. My first real close call or incident I had was actually getting into the Vortex Ring State (VRS).

It was thirty years ago when I was doing aerial work in the Swiss Alps and the helicopter developed VRS. For some reason the VRS stopped suddenly, just before the trees, and I didn't know why. I flew home with shaking knees and didn't sleep for two nights afterwards because I realised I could have easily died. So over those few days and nights I developed this new manoeuvre now called the Vuichard Recovery Technique to get out of the Vortex Ring State. I realised that the tail rotor was still working under normal conditions. So why not use the tail rotor thrust by increasing collective to maintain the heading. Simultaneously apply opposite cyclic (15°-20° bank), cross control to get a lateral movement, this will move the helicopter laterally and move the rotor in the upwind part of the vortex and helicopter stops immediately the huge rate of descent and the VRS recovery is completed.

I went back to the aircraft, put it into a VRS, and I used the technique. I got out of the fully developed Vortex Ring State in only 20 feet. Sadly nobody believed me over the next twenty years because even though I was a flight instructor I really had no network, until 2011 where I met Tim Tucker.

Tim is the safety pilot for Robinson Helicopters and he was doing a safety course in Switzerland. I asked him if I could show it to him. At first he was not very happy because he did not want to break his neck in a foreign country but after I had the opportunity to show him the technique in flight, he was amazed and he asked if he could try. He he was able to get out of a fully developed Vortex Ring State in 30 feet. He said, "Claude, I have never been amazed like this in forty-five years of aviation." He wanted me to provide him with the technique and information. Robinson Company conducted flight tests at the factory with a test pilot, and a handful of experienced Robinson factory flight instructors tried it out in the network. Everybody came back saying, "Your guy in Switzerland is right." Robinson implemented it and put my name on it. In 2018 I got four reports of pilots using the technique and saving the lives of everybody in the helicopter.

This gives me the energy to load my batteries to deliver safety seminars and flight training around the world. I have a clear vision to achieve zero Vortex Ring State accidents in the helicopter industry.

If people listen to what I am saying on how to get out, and also on how to avoid it, I am convinced that we will get to zero.

I used the technique in 1999 while night flying for the Swiss government searching for a missing aircraft. I was operating with night-vision googles and suddenly we went into the vortex - I took control immediately as I saw the rocks moving very fast in front of me and I looked on the vertical speed indicator. I did my manoeuvre in an instant as it was now a reflex, we pulled out at around 100 feet altitude. So I saved my and the crews lives once more using the technique.

I have flown more than twenty types of helicopter, from the Robinson R22, R44, R66 the Bell 206, the AS350 series, AS365 Dauphin, AW109, AW119, AW139, almost everything, and the technique works in every helicopter with an anti-torque system

In the mountains, the nicest helicopter is the SA315 Lama, despite being designed fifty years ago. I still consider it the nicest aircraft, great visibility. I like to fly most aircraft. I do like the H125, also a nice aircraft and I am looking forward to the Swiss made Kopter, being a new aircraft made from scratch as opposed to being from previous designs and models.

Sadly, I have lost around forty friends over the years. What is amazing is that over these years there have been many changes and new rules implemented but we have not changed the number of accidents. The only way to change the number of accidents is to work on the human factor issues — we can eliminate a large number of accidents by teaching and using better techniques, such as the Vortex Ring State technique, autorotation techniques, quick stop techniques to brake the aircraft, avoiding entry into IMC.

My non-profit organization - the *Vuichard Recovery Aviation*

Safety Foundation, was created to help train and avoid these ongoing accidents. My vision is to reduce the human accident rate in the helicopter by 80%, that means globally a reduction of the helicopter accident by more than 60%.

I have saved hundreds of people and I look forward to saving hundreds more with my techniques. It is an amazing feeling when pilots send me messages that they have saved their own lives and the lives of others because of the technique that saved my life.

I am convinced that I am on the right course around the world to save a lot of human lives.

Learn my technique and fly safe.

Claude Vuichard

A former line pilot and flight instructor who developed the Vuichard Recovery technique. Receiver of the Salute to Excellence BLR Aerospace Safety Award - HAI 2018.

INTRODUCTION
MARK OGDEN

"Helicopters are renowned for working in some of the most remote and inaccessible locations on the planet."
HeliOps International Magazine

I first flew with my father when he worked for George Cadet in Brisbane in the early 1960s. I was seven years old and got the bug. Then, as an Air Force Cadet I flew gliders out of Warwick in Queensland, Australia and managed to score a Royal Australian Air Force Scholarship to have my pilot's licence before I turned 17. I could fly before I could drive.

Wanting to be a military pilot, I tried joining the Royal Australian Air Force (RAAF) but I was a year too young. Walking downstairs at the recruiting office, I saw a photo of the Navy's aircraft carrier and discovered their recruiting age was a year younger than the RAAF; so I ended wearing naval wings. Their Pilot's Course in those days was with the Air Force, flying the then new CT-4

Airtrainer, the Macchi Jet and finally landing on the UH-1 heli-copter. I didn't want to fly helicopters, but when I got to 5 Squadron (RAAF) to fly the Huey in 1977, most of the instructors had come through Vietnam and there was so much to learn from them. Flying helicopters was the best flying, and I had found my niche. If one could not hover with a tree limb on your chin window, then one was not meant to fly helicopters.

When I returned to the Navy, I flew the UH-1 before moving to the venerable Sea King. Having survived a bad accident in a Sea King in 1979 on my first day at sea, the rest of my flying career went a little more smoothly. In 1981 came my instructors' course with the RAF where I flew the Gazelle, then back to the RAAF to teach. Later, I managed to get onto the best project a pilot could hope for, bringing the S-70B-2 Sikorsky Sea Hawk online. After spending some time in the USA on the SH-60B American Sea Hawk, I was back in Australia flying the Australian version which I continued until the end of my naval career. I still think the Sea Hawk is probably the best helicopter in the world — built like a brick, with a lot of power and just a joy to be in.

The last course I did in the Navy was the Accident Investigation Course at Cranfield University in the United Kingdom, adding crash investigation to my resume.

After leaving the Navy in 1996, I spent a few years with the Bureau of Air Safety Investigation (BASI) as a crash investigator and still occasionally flew, most GA fixed wing but for me, the flying just did not present the challenges of naval aviation where I found the flying from various ship decks in different weather, interesting missions and having a flight crew that depended on one another was vastly different from civilian flying. Being part of a larger operation provided a huge sense of satisfaction. But don't get me wrong, as you will read things can get messy in any model of helicopter or any form of flying. BASI provided another piece of an aviation career and I learned as much there as I did in the Navy and rounded out my atti-tudes and knowledge. It was here I truly learned about human performance and started to understand organizational issues and their impact on aviation operations.

BASI became the Australian Safety Transport Bureau (ATSB) and I looked after the Perth Office from 1997 through to 2002, before returning to the Navy. Then I became involved in aviation accident investigation for an oil and gas company. With knowledge acquired over the years, from operations to investigations, I have been helping manage the company's aviation safety standards and operations in places such as Papua New Guinea and Alaska. While not being so much a pilot these days, the experience and knowledge I gained through the years in my various roles has proved essential to let me do what I do now.

All that preamble sets the tone for what I believe contributes to safe flying. There are three rules to keep aviation safe:

1. Always be aware — where are you? What is happening around you? Project that forward to understand what will likely be happening ahead of time. Listening to the engines, the transmission and knowing what the aircraft is doing are all part of being aware. Managing the energy in your aircraft is also an awareness concept – know how to manage the energy inherent in a flying aircraft. Being fully aware is essential for both air and ground crew.

2. Understanding Risk — is the planned flight worth the risk? Risking your life is one thing, but don't forget lives of others are your responsibility; whether that's the crew, passengers or those who may be left behind. The aircraft is in your hands and that is a significant responsibility not to be taken lightly.

3. Essential and basic flying skills must be maintained—
    without those, there's nothing to fall back on when (and it
    will) the automation fails.

If you have the above and have learned to fly a helicopter well,
you could almost fly anything.

I found the skill levels differed greatly from fixed wing. I know
that every time I got in a helicopter it gave me more joy than a fixed
wing aircraft. I believe it also made me a better thinking pilot. I am
not saying fixed wing is not challenging, I just found flying helicopters more challenging than any other flying I ever did.

As a crash investigator it seemed that most helicopter accidents
occurred in the agricultural industry (including mustering). Many
pilots lacked a full understanding of their aircraft. Aircraft were often
overweight and pilots often over boosted (especially in the Robinsons)
or tried flying with degraded rotor RPM — or they'd tried incorrectly
maneuvering the aircraft. It seemed, in the mustering industry for
example, pilots were musterers first and pilots second. Basic skills and
knowledge were lacking. You just don't often see so many similar
mistakes in other areas of the helicopter industry.

Many of the non-mustering accidents were wire strikes. In most
wire strikes, many pilots knew the wire was there but had forgotten.
Often the accident was late in the day so fatigue likely played a role –
degrading a pilot's situational awareness.

Always wondering what is happening and examining the *what
if's*... "expect the unexpected". I found with the Sea King, pilots had
close calls to reflect on almost every flight. I say that somewhat in jest,
but the Sea King was a 1960s helicopter; there were no self-checking
systems and we always expected things to go wrong (and they often
did). That was just part of flying this model of helicopter in that environment (low level over water at night). The change to the Sea Hawk
was a jump of two or three generations of helicopter. The Australian
Sea Hawk was extra-reliable and super-powerful, built like a brick
and very forgiving. Despite 30 years of operations in some of the most

unforgiving conditions you could throw at an operation, the Squadron did not lose a machine and indeed in 2018, the Squadron managed to 'pay off' the same helicopters that were purchased all those years ago mostly because of the investment in the initial and the ongoing training for both air- and maintenance crews.

The lack of basic skills in aviation is showing up now in the airline industry. Many of the latest accidents supports this thinking. Pilots are forgetting the basics, such as the old adage, "power plus attitude equals performance", or "Aviate, Navigate, Communicate". Having said that, we also see occurrences that would have been significant accidents if the crew were not on top of the game.

The Qantas Flight 32 and CACTUS 1549 could have been serious accidents, however both crews on the flight deck had the skills, situational awareness and knowledge to get the aircraft down safely. Any failure to invest in long term training is almost a guarantee of failure and accidents.

Any close call I had, I made sure I tried to learn the lessons. I made many mistakes but survived to tell the tale and to pass that knowledge and experience on through the later stages of my careers. In this industry we need to ensure our pilots learn from every mistake. If any pilot denies their responsibility or accountability, and they think they are untouchable, then they are heading towards an accident. Believe that you are fallible and you will make mistakes but ensure you learn from it with the personal responsibility being with you and not others. Learn from others and always remember, "there by the grace of God, go I".

Many times I have said to myself, *wow I didn't think of that...* and that's where I argue the great saying of *there are old pilots and bold pilots but few old, bold pilots.* Always remember that as a senior pilot, you will be emulated and held to be an example. I was the guy who thought I could fly lower and faster than anyone. I got away with it, not sure though if it was due to superior skill or pure good fortune. My responsibility was bought home one day when a senior Squadron pilot pulled me aside and said, "Remember other young pilots look up to you and they will emulate you". He asked if I wanted the responsibility for a young pilot trying one of my manoeuvres and killing themselves? I realised then the responsibility of senior pilots.

The valuable lesson I learned was that not for me but for the aviation industry in general; information and techniques are passed from pilot to pilot, usually older to younger. Pilots teach other pilots.

Mark Ogden
Editor HeliOps

Australian ATP Licence (Rotary) with endorsements and CPL (Fixed). 5000 hours experience in offshore and onshore ops commercial, military, firefighting and SAR. 3000 hours instruction. Air Safety Investigator, published manuals, policies and procedures. Presently working in aviation management.

PROLOGUE
FLETCHER MCKENZIE

"Most accidents originate in actions committed by reasonable, rational individuals who were acting to achieve an assigned task in what they perceived to be a responsible and professional manner."

Peter Harle
Director of Accident Prevention
Transportation Safety Board of Canada

EVER SINCE I WAS YOUNG, I've marvelled at all things that fly — from paper planes to the early aircraft prototypes to the massive Spruce Goose. My father passed his passion for aviation on to me. He was a man who loved reading, learning, and talking about aircraft. Before he died, he was following the developments on the 787 and the progress of the A380, and we had many discussions around the classic argument of who is better - Boeing or Airbus.

In his early years, my father worked for the National Airways Corporation (NAC). Whenever we drove past the Kaimai Ranges, he would talk about the DC-3 that crashed there in 1963 and how he'd known one of the flight attendants. I still have the DC-3 model he made for his office, hand-painted in the NAC colours. I'd pulled it apart as a child but its battered remains live on in my own office. I always wondered how and why that DC-3 had crashed, and it wasn't until years later when I read Rev. Dr Richard Waugh's book on the accident, that I knew what had happened.

I've detailed some of my personal experiences with people and flying. My journey into the skies has been different to that of most pilots, and I've come into the industry through a peculiar series of chance encounters and personal friendships. It's a journey filled with learning from so many pilots — a constant number of lessons from the sky, and the realisation that the reflection and time taken afterwards is where the learning starts. Some of those friends have died in air accidents. If what I've documented in these pages saves just one life, then my work in publishing these books has been worthwhile.

This is my fourth book in the series. I have read thousands of lessons, and these lessons have had a hung impact on my flying habits. I wanted to show how powerful these stories have been for me.

Some years ago I read a story in Vector magazine (New Zealand Civil Aviation Authority - CAA) about an incident with a cockpit seat where the pilot pulled back the control column on take off and the aircraft seat moved aft. With the pilot holding onto the control column, they pulled up as they slid back, forcing the aircraft to jerk upwards into a stall. Somehow the pilot regained control and the aircraft landed safely without further incident. The chair rail lock had failed.

I remember hoping it wouldn't happen to me. Interestingly, despite reading the article, I never checked the seat structure or asked anyone about any modifications made to the aircraft I usually fly. However, what I did do after reading the article was, without fail, during my preflight checks, I always gave the seat a good shake to ensure it was steady and the locking mechanism worked. And once I was in the aircraft and buckled in I always shook the hell out of the chair to ensure it wouldn't move on take off. However, I never examined the chair locking system or tried to understand the seat structure in the said aircraft. I assumed that the owner and our maintenance engineer would have checked this.

Reading a story from the UK's Confidential Human Factors Incident Reporting Programme (CHIRP) — "SEAT FAILURE ON TAKE-OFF" under the Airworthiness & Maintenance in Chapter 1 in *81 Lessons From The Sky* (*General Aviation*), got me wondering how many chair legs are locked in place. I assumed that all four were locked in. The next time I was at the airfield I checked, and found that none of the back legs were locked and, to my amazement, just one leg on the inside was used to lock the chair in place.

One 5mm diameter steel rod was holding my 95kg frame into place. When the aircraft was designed in the 1960s, the average pilot weighed around 65-75kgs, if only that were still the case. I couldn't believe I never realised that only one seat leg was locked. A friend suggested I design and manufacture a device for this. First, I searched the Internet for a seat locking devices to see if any were already on the market, and found a few designed for various seat locking issues. I have now purchased a locking device for my seat. The device locks onto the rails and it acts as a stopper if the seats moves back.

Since the release of *81 Lessons From The Sky* (*General Aviation*), I have had various conversations with pilots about their seats, hearing about other pilots who went through similar occurrences and, who through sheer luck, are still here to tell their story. I recently flew a Cessna 172S (injected) for my BFR (Biennial Flight Review). During my preflight I checked the seat mechanism and the seat design. It is noticeably different from the aircraft I usually fly, with four horizontal locking pins on each side as opposed to one. It also had the tensioner (a Cessna modification still available at the time of writing) to ensure the seat will catch if the pins fail.

The seat lesson is just one of the dozens and dozens of stories that made me think further about what I do, or what I need to do, to miti-gate the risk before or during my flying.

As a teenager in the Air Training Corps, I was selected for a gliding camp at RNZAF Base Hobsonville, where I learnt the theory of flight, going solo at the age of 16. The youngest instructor on that course was 37, and had a number of flying awards to his name. Some months after my course, he was killed in a glider crash. To a teenage boy, it was impossible to understand how that could happen to an experienced pilot.

In 1990, one of the requirements of my Bursary Maths paper was to complete a statistical project. I chose a subject I was interested in — aviation, specifically **Aircraft Accidents 1979 to 1989.** I had to find the relevant statistical information and build out theories explaining what the numbers suggested. I have constantly referred back to those findings as accidents occurred in the aviation industry, to see if my hypothesis was correct, given the circumstances of the given incident. The correlation has been interesting.

Safety was paramount when I flew gliders as an Air Training Corps cadet, but it wasn't until I began my Private Pilot Licence (PPL) training, that I began to understand the factors which would lead to an accident. *Human Factors* was the most thought provoking book I had read for a number of years. The deeper I read into the situations pilots got into, the more I understood the factors which can lead to poor decision making.

After several years of building an advertising agency, I created a business around my passion — to do something that got me closer to aircraft. In 2010, I began production of the television show *Flight-PathTV* - a magazine style television show on aviation. *FlightPathTV* was on air in sixty-one countries (including the Discovery Channel with Rugby All Black Captain Richie McCaw), and has been trans-lated into various languages including Mandarin.

Over a period of eighteen months, we interviewed around one hundred pilots from around the world, and spent hours listening to personal stories from pilots on what inspired them and how they become pilots. The first time I heard that one of our interviewed pilots had been killed in an aviation accident crushed me. The pilots we interviewed were incredibly experienced, and the news of his death was inconceivable. Over the last seven years, this toll has risen to six pilots, with even more involved in non-critical accidents. Why is this number so high?

As part of my ongoing role in filming aviation stories, I meet and interview pilots from around the world to find out what they read and what inspires them to be safer pilots. What did they do? How did they do it? And how have they changed their processes to become safer in the air?

Being a private pilot and having a young family, I read every aviation safety magazine and numerous books on flying to learn from those incidents, especially the near miss stories to ensure I don't make the same mistakes. Through other pilots sharing their stories, I become a safer pilot.

I get to work with a number of entrepreneurs from around the world, through EO (Entrepreneurs Organisation), leading them in strategy planning and training them to experience share between each other and to learn from mistakes — the good and the bad. This is a proven process I want to add to the aviation community.

EO is a global, peer-to-peer network of more than 14,000 influential business owners in 74 countries, and they employ a unique communication model which provides unparalleled access to the wisdom of your peers during confidential monthly meetings. It is called 'Forum'.

Forum came out of the desire to have a safe environment to share and learn from others' experiences. Extensive research was undertaken to develop the concept. Building from early small group theory, the key objective was to create a supportive environment for members without fear of confidentiality being broken, and without risk of being judged by others, to share, learn and grow within a close group of peers. The language protocol supports the risk being taken by others in Forum, and is what makes Forum a safe place. The

"Speak only from Experience" mindset encourages people to find their own answers.

In my line of work, I talk about helicopters and associated products almost every day; they are exciting, complex flying machines and can be challenging. These machines add a lot of value to people's lives globally.

The first helicopter I ever sat in was a Royal New Zealand Air Force (RNZAF) Bell UH-1 Iroquois; that Huey lead to a lifelong fascination with helicopters and being a cadet in the Air Training Corps in the 1980s gave me amazing opportunities. Since then I've been lucky to fly as a passenger in several civilian helicopters around the world, with various military units on exercise and occasionally fly one.

Every time I've sat in a cockpit, and especially since I started flying different aircraft and attaining different ratings, I've learnt new lessons.

As a sixteen-year-old I learnt to fly a Blanik L-13 glider and grappled with understanding how the aircraft performed — I didn't even have my driver's licence at that stage, and here I was in the air.

Fast forward several years, and I found myself trying to understand the flight characteristics and feeling of the stall buffet of the Cessna 172 when doing a streamer cut in the fastest possible time for the New Zealand Flying National Competition. Onto realising how fast I had to react if we lost power in a twin engined aircraft upon take-off and how quickly we needed to react because of the asymmetric effect.

The ongoing constant monitoring and recognising, at low levels, the glider behind me and if it was going to get out of position before it was too late to pull the tow line.

To flying an aerobatic sequence in the "1000 metre box" at my first aerobatic competition without a safety pilot. Understanding how and what the aircraft was doing or sometimes not doing because of my input, coupled with knowing exactly where I was in terms of space and ground reference and keeping a close eye on my altimeter for height below and above.

Monitoring the situational awareness and knowing what the correct reaction will be before something happens is essential. Situational awareness can be taught, but it comes with experience. There is no shortcut.

It has been a journey of learning with so many different pilot types mixed in — a constant number of lessons from the sky, but more importantly, realising that the reflection and time taken afterwards is where the learning starts. A number of my non-aviation friends think I am crazy. In reality, I am just overly passionate about aviation. The feeling of freedom is incredible and I couldn't get the same excitement and adrenaline anywhere else.

Remember that Bursary Maths assignment back in 1990? I was studying Math With Statistics, Maths With Calculus, and Physics so I could join the Royal New Zealand Air Force. Which was why choosing aviation as my subject matter was so easy. As I researched aircraft accidents in New Zealand, without the benefit of the internet, I was forced to pick up the phone to call the office of the New Zealand Civil Aviation Authority (CAA). From that phone call, I discovered I could purchase journals covering all the aviation accidents in New Zealand. I bought eleven of those journals.

Suddenly I was reading about aircraft incidents, and at times fatal crashes. As a teenager, it had an effect on me. The assignment was completed, and there were some interesting results. But now that I am part of the aviation industry, the numbers are more than just statistics published in a journal. They are someone's life tragically cut short, someone's mother or father, a son or a daughter. That is the hard part of statistics — when they are numbers on a page there is just no emotion involved.

I had planned to republish that research paper, and incorporate the latest statistics, but I was not 100% sure how the report and its findings would help the industry. Whilst thinking about publishing, and the desired outcome, I started writing a few stories myself, the first one after a personal air incident. I was encouraged by a good friend and instructor, to write about that incident, to understand the learnings from the event. And that is how these books began.

During my time working within the aviation sector, firstly with the production of *FlightPathTV*, and manufacturing, selling and buying military and helicopter parts, I have been able to meet and speak with many influential aviation people.

One meeting was thanks to Rob Neil from *Pacific Wings Magazine*, where we interviewed Lynn Tilton, the CEO of *MD Helicopters* during her trip to New Zealand in 2007. The interview included Joe Faram with his MD520 NOTAR (no tail rotor), and Lynn shared her insights to how she revived a legendary helicopter business. We flew onto Oceania, an MD Service centre. Years later, I worked with *MD Helicopters* — flying and filming MD500's and working on strategy and tactics thanks to Oceania and the MRO (Maintenance Repair Overhaul) software provider Aeronet with Aaron Shipman. I have been lucky enough to jump into aircraft piloted by Stephen Boyce. Always memorable — from an ex Japanese Coast Guard Iroquois doing a blade timing test to flying Helisika's MD530F and JB's 500 and for us to film footage for an *MD Helicopters* launch project. This ongoing work in the MRO arena gave me an understanding for the parts and rotables supply, and this led me to the space where I am today — representing the Asia-Pacific region, Africa and the Middle East working with an overseas based supply company. We also licensed a loadspreader floor design for the RNZAF's AW109 and NH90 helicopters and brought in high tech manufacturing companies to produce a very light armour floor to protect people and the asset.

I interviewed Frank Robinson, discussing aviation in the 60s and 70s, and how he wanted a cost-effective helicopter, and thus the R22 was developed. I met with his son, Kurt Robinson, who now leads the Robinson helicopter company.

Lisa "Choppy" Paterson in Queenstown flew us to the highest golf tee in the world — on the Remarkables mountain range in

Queenstown. She then dropped us off at Michael Hill's world renowned golf course. Choppy started a company to build a helicopter black box after her son, and her companys' chief pilot, died in an R22 crash.

We flew with Phill Hooker from Tauranga, New Zealand who operated a fixed wing training school but also had several helicopters — a Bell 47 in M.A.S.H. (Mobile Army Surgical Hospital) colours, a Kawasaki 369 (a H500C made under license) painted in U.S. OH-6 livery from the Vietnam war, and an original Bell 47J Ranger — once operated in the oil and gas industry by HNZ in New Zealand. It also flew Elvis for one of his movies. The Ranger was one of the first helicopters Phill flew, many decades later he uncovered it in Queensland Australia and had it rebuilt. The 47J Ranger is based on the Bell 47 but with many refinements. With the restoration project completed, it's a great example, painted in a bright orange colour. We flew both helicopters to film a sequence with the H500C and troops on the ground in Tauranga for a Vietnam reenactment for the Tauranga City airshow. When we came to shoot the airshow, I tasked one of our cameramen to film the airshow from the Ranger, but he said he wasn't comfortable. He confided that another cameraman, a close friend, had died in a helicopter crash made famous by having a well-known media personality on board, with the survivors swimming to shore.

In Cape Town, South Africa I flew in a Vietnam Vet UH-1. Pushing the limits of the helicopter packed with a group of colleagues was the one experience that made me think about safety and margins. By this stage I'd started to understand the limitations of helicopters and pilots, and I had been building up a knowledge of how air forces operated (with built-in margins for error). While the Cape Town experience was exciting, there was little margin for error as we flew towards trees and other local fauna. Speaking with the pilot afterwards, it transpired that he was an experienced crop-duster pilot with around 20,000 hours. Through experience, the pilot has honed his own margin of error. If only I'd thought to ask him what lessons he'd learnt.

The great aviation photographer Gavin Conroy, organised for us to go up in a Bell 206 to film Sir Peter Jackson's World War I reproductions — a Fokker, and a Sopwith Camel. We recorded some incredible footage with a great pilot capable of holding the helicopter in a very stable position.

Our production company filmed the New Zealand Helicopter Nationals (thanks to Roy Crane), which included the RNZAF with their dated but proven 50-year-old Bell UH-1, R44's, Squirrels, R22's. Highlights included filming pilots opening beer bottles with the skid of their machines. Jumping in the back of a Huey for the navigation task flying with doors open and filming the crew operating as a team. Lastly filming Pauanui's famed property developer Leigh Hopper in a race between a helicopter flown by him and his son Stacey in a Subaru rally car. The helicopter won. Stacey was tragically killed flying a Cessna 206 a few years later.

Not content with filming in New Zealand, we set off to film the various Everest rescue helicopter companies in Nepal, filming emergency helicopters rescuing trekkers and high altitude climbers in Nepal during the climbing season. Armed with expensive permits to film, we recorded several stories for the second series of *FlightPathTV*, capturing tourist locations, local villages and interviewing the staff and air crew in Kathmandu. We filmed the crew and the flight flying into Lukla (one of the most dangerous airstrips in the world — depending on who you ask) on the DH4 Otter, interviewed CEOs and also did the Everest Scenic Flight. Being embedded with the rescue helicopter crews in Lukla Nepal fed us full of adrenaline. With no fuel available at Lukla (or anywhere else in the vast area), the helicopters have to fuel up at Kathmandu airport. Once in Lukla, the engineers use hoses to suck (via mouth) the JetA1 out of the helicopter tanks into plastic fuel (Jerry) cans which are then dropped at various points along the way to the evacuation point, this is to keep the helicopter as light as possible in order for high altitude operations. Once the patient is on board the heli-copter refuels back at each stop with the fuel it had just dropped off — the challenge is ensuring it has enough fuel for each leg without any weather diversions. It came down to how high and how far the trip was from Lukla and then how many stops they had to fuel up at and understanding the terrain and weather that can change very fast. Sadly there have been numerous fixed and rotary wing crashes in Nepal, involving some of the crew we filmed.

I met the passionate professional photographer and media personality Ned Dawson, owner of the magazines *HeliOps*, *Air Attack* and *HeliOps Frontline*. Over many years he has helped connect us to various organisations. While visiting Hong Kong, Ned connected us with *Heliservices (HK) Ltd* working from various spots which included a heliport in downtown Hong Kong. Few people get to see the exclusive China Clipper lounge tucked away high upon the 30th story of Hong Kong's grand Peninsula Hotel, one of the city's most elegant art deco landmarks. Ned suggested we should talk with *HeliOps* editor and helicopter pilot Mark Ogden, who wrote the introduction for *71 Lessons From The Sky*.

Working in the industry I have attended the HAI (Helicopter Association International) Heli-Expo several times now and at the last HAI I met Claude Vuichard, an experienced mountain rescue pilot based out of Switzerland. I interviewed Claude about his Vortex Ring State (VRS) recovery technique and his non profit organisation helping train pilots around the world. Tim Tucker, the Robinson Helicopter safety pilot, was the one who supported Claude on his journey to save the lives of pilots. Chapter 4 covers the Vuichard Recovery Technique.

During one airshow we interviewed pilot Roger Buis, who has amassed over 18,000 hours since 1980. He flew a low-level choreo-graphed performance set to music, which included playing with his Yo-Yo and racing around and picking up barrels. The helicopter he flew was Otto, a Schweizer 300C, he also flew during the night show — a spectacular aerial ballet with the fully articulated, three-bladed rotor system, making it ideally suited for sensational rotary displays. One of Roger's many amazing stories was about an autorotation at night when he lost an engine with a student, aiming for the only light visible they landed and after climbing out of the helicopter they soon realised it was on the top of the cliff — only a few more feet and there would have been a different outcome to the story.

I also talked with a number of good friends, one experienced pilot Stephen Boyce who flies various flying machines around the globe. Stephen's thoughts were, "the limitation section of the Flight Manual must be treated as gospel — it must be read and fully understood before flying." Another friend, business associate and UK pilot now based in the USA, Ian Dodds said, "Helicopters and golf courses, on three occasions I have come to realize that helicopters become magnetic to golf balls when parked near the club house, maybe it's misadventure or just poor planning. An old military pilot once told me the only way I would improve is to keep training, 19 years on I still do."

Filming in the South Island of New Zealand we met Sir Tim Wallis several times and filmed him in front of his old Mark XIV Spitfire. Sir Tim is a famous deer recovery pioneer from the high mountains of New Zealand. We filmed other helicopter legends, Keith McKenzie and Bill Reid. We met Bill during the Omaka Classic Fighters Airshow, where we filmed his immaculately restored 1937 WWII Avro Anson.

After a number of interviews, it made sense to talk further with these experienced helicopter pilots for this book, so I reached out to Kiwi helicopter legends Keith McKenzie and Bill Reid, who have over 40,000 flying hours between them. It has been a privilege talking with them about the aviation industry, and to fly with them occasionally. I've included stories from both Keith and Bill below. I would like to thank them for their willingness to share their experience and wisdom with you.

Bill Reid

One of Bill Reid's first lessons occurred shortly after he had received his private pilots licence, in the early 1970s when he rented an under-powered Cherokee to fly his girlfriend a short twenty-five minute trip from Nelson to Omaka in the South Island of New Zealand. There was a large Cumulus cloud in the way which he thought he would lightly brush through the top of and be out of in a few seconds. He got in deep, becoming disorientated and ended up in a spiral dive. By sheer luck he came out of it before they hit the ground, and getting away with it without tearing the wings off. Bill decided after that to get a lot more instrument training.

In 1980, in the days before dedicated rescue helicopters, Bill Reid was working for his fathers company *Helicopters New Zealand* when he received a midnight phone call from Peter Tait the company's General Manager saying the police were requesting a medical evacuation from a road accident in Takaka. They only had a couple of Bell 47's available one of which had recently been converted to a Soloy 47 with an Allison 250 turbine greatly improving its performance.

They had a stretcher made up that protruded all the way forward from the right hand seat into the front of the big bubble of the 47. The doctor or paramedic was then able to sit in the centre seat and administer to the patient. Heading off they flew to the accident site surrounded by low cloud and mist. After picking up the critically injured motorcyclist they headed back to Nelson hospital. The weather was improving and the night was light enough now with glimpses of the moon appearing. Bill thought he was flying at 1200 feet just off the coast of Rabbit Island when he saw something shimmering in front of him which was sort of mesmerising and distracting. Then the little sliding side window blew out which woke him up and he realised that he was at almost the VNE (Never Exceed Speed) of the aircraft which was normally almost impossible to achieve in level flight but with the Soloy conversion was fairly easy. Then he saw

more shimmering and noticed he'd misread the altimeter he wasn't at
1200, he was only at 200 feet and in a shallow dive towards the sea.

After composing himself and climbing back to a safe height,
memories of his earlier disorientation experience came back and he
wished he had been in a machine with more instruments to refer to.
The 47's only flight instruments were an airspeed indicator and
altimeter. However luck was on his side, and they got the patient to
the hospital in time. A good result in the end.

The lesson here being always double check what basic instruments you do have, maintain your situational awareness and do not
become complacent, suffering from 'get home itis' as you get near to
your home base or airfield. Bill has lost a number of friends who have
crashed within sight of their home base.

Bill's latest lesson is very recent, and could have ended very
differently. The lesson is so worthwhile, I asked him to retell it here
in his own words:

It was Day 13 at New Zealand's biggest forest fire since the 1950s
with the largest fleet of aircraft ever assembled at one fire in that
country. 22 helicopters and 2 fixed wing were working the fire on
several fronts in the pigeon Valley area south of Nelson.

On the afternoon of the accident, the main fire was under
control and we were dampening down the hot spots. I was picking
up the water from the dipping pond with the monsoon bucket on a
long line and I remember looking down and the bucket around the
lip didn't look quite right, but it was difficult to see looking vertically down on it and I thought maybe it didn't have enough water
and dipped it again and it still looked just a little bit strange but the
indication on the load cell was that I had a good load and a BK117
was behind me waiting to dip, so I moved on and went up and
dropped the load on the fire and turned back to the pond.

As I accelerated through about 70 knots and at about three
hundred feet off the ground I got this almighty yaw — a real big
sideways push, which I thought was wind shear turbulence as we

often get that in this area but when I got another one from the other side that really made me sit up. Next there was an almighty bang and shaking and the helicopter was doing some fairly wild oscillations I thought, "Jeeze Bill, I got to put this machine on the ground". I managed to get a mayday call out as I knew I was really in serious trouble although I still did not really know what was happening and actually did not realise until later that the fire bucket had flown up and hit the main rotor blades and it's long line had become hooked over the horizontal stabiliser subsequently entangling in the tail rotor output shaft which then winched the bucket into the tail rotor blades and severed one of them.

I saw a skid site with a fire truck and crew in front of me and I thought I would head towards them but I found out very quickly that I couldn't steer the helicopter, it was locked into a gentle left hand turn.

I had jettisoned the bucket as I did the radio call and that had helped calm some of the wild feedback I had been getting through the cyclic caused by the bucket now trailing out behind me still over the stabiliser and causing an extreme aft C of G.

My daughter-in-law Rachael was flying another Squirrel with the fire boss on board not far from my position directing all the aircraft and she heard my distress call and quickly turned towards me. As I slowed down, the aircraft started to spin with no tail rotor, so I closed the throttle and shut down the engine, and autorotated into some burnt trees ending up crashing nose down into a small bank, I literally shut my eyes in the last couple of seconds as I thought this was the end. As I opened them I saw what I thought was smoke and thought oh great I have survived the crash but now I am on fire. I started to undo my harness to get out but realised my right ankle was looking a bit floppy. I had been pushing so hard on the right pedal trying to get the helicopter to turn that way that when I hit the ground, all my weight shot forward and down through that leg and I had bent the stainless steel pedal and broken my ankle.

I soon realised I was not actually on fire, what I thought was smoke turned out to be fine ash. I had come down in an area that the fire had been through the day before and Rachael who by now had arrived at the scene was hovering near by to see if I was still alive was whipping up a cloud of dust.

I managed to put out a call on the radio that I was okay before switching off the electrics. The fire crew that I was trying to get too were soon on the scene and helped me out of the helicopter and into an ambulance.

Our initial investigation of the incident found that there was a fault with the fire bucket — the stainless steel ring that gives the bucket its shape around the top had broken free from the Velcro fasteners that hold it in place meaning the bucket lost its aerodynamic shape and had become totally unstable like a large sail or kite flying wildly about in all directions.

The lesson here was the strange shape I had seen on that last fill should have alerted me to stop and find out what it was I saw but I had never heard of the problem before and was not expecting any such violent reaction. Since then, other operators have come forward stating it had happened to them, but without the same consequences as they were fortunately going slower at the time of failure.

Some people have said to me why did I not I just let the water out and fly back to the staging point when I noticed the anomaly with the top of the bucket but the thing to remember is that although I did not know it at the time whilst there was water in the bucket it was safe. Without that weight holding it down it became a lethal weapon and the situation could have been a lot worse if I had emptied it and the bucket had gone crazy when I was flying back over a populated area. In hindsight I would have put the bucket down as close to the pond where I could land and sorted it out there.

Keith McKenzie

Filming *FlightPathTV* was when I really started working closely with helicopters, and my first experience was with Keith McKenzie's Hughes 500D at the Ohakea RNZAF open day in 2007. We rigged an ATAK mini camera on the tail — our first lesson in understanding what forces play on the aircraft, the control surfaces, where engine heat exhaust is, what and where the display will be on the airfield and where will the sun be (when the actual display is operated).

We worked with Keith and his son Scott (then an instructor with the RNZAF) and an engineer to ensure they were happy with the installation. As the handling display got under way we waited patiently to see what awesome footage we'd secured. Sadly the camera randomly stopped working as did many of our cameras over the years. Meeting Keith and Scott was insightful. Their family has operated aircraft for decades, from the deer recovery industry in the same Hughes 500D since 1980, to operating Britten-Norman BN-2 Islander aircraft and flying passengers to Great Barrier Island, to operating aircraft in the Tongariro mountain area.

Keith McKenzie's most memorable flying moments were touching the skid of a Bell 47 on top of New Zealand's highest peak of Mt Cook while deer hunting in the 1970s, landing his Hughes 500D in the crater lake basin on Mt. Ruapehu after the volcanic eruption. The lake was empty after the eruption and lahars, but the mountain was still active and steamy, and he was tasked with dropping in scientists and their gear to monitor the activity. He sold the aircraft in 2015 when he finally retired from helicopter flying.

Keith is a lucky man — walking away from a number of wire strikes. Chapter 3 details five wire strike stories. It was only after listening to Keith's stories that I decided to include them, as he went on to talk about training.

Keith has shared his experiences with other pilots, but I have been the first to interview and write about those wire strikes. Keith's first close call was fairly early in his career when he hit wires in Australia. He was in the Northern Territory, Tennant Creek ferrying a Bell 47 and he needed to land as he was desperate for a toilet stop after flying for over five hours.

He came into land and never saw the wires he hit. He released his seatbelt and bolted through what was left of the large bubble canopy and leapt across the large ditch that the water main had been dug for and ran up the other side and into the scrub - the helicopter totally destroyed. He recalled coming back to the site for the helicopter the next day and seeing the distance he jumped across, calculating he would have won a medal at the Olympics for the long jump. It is amazing to see how adrenaline helps us achieve things we usually couldn't do physically.

The lesson for himself was that wires are probably the most terrifying thing you can encounter while flying at low level. In the agricultural sector it is always a risk, every time you go onto a new property, even though you ask the owner where the wires are — sometimes they forget about the odd one or new one. They may recall the location of the nine places with wires but accidentally forget about the tenth one.

The following story and thoughts I have left in the first person, so Keith is retelling the story below with his thoughts and comments.

My other scary wire strike was on a Saturday morning in Auckland, New Zealand when I hit a 33kv power line.

I was flying to a site in the Waitakere Ranges flying gear in for a residential house project, I was descending through the valley and at the last second I saw wires, I reefed the machine back on its tail, the wires hit on the nose and they arced with sparks everywhere, one of the wires broke and wrapped itself around one of the skids, we came to a grinding halt and gradually started to climb.

I had the power in and suddenly the machine started to tip over so I slowly bought the machine back down, untangled the skid and landed close to a house in the valley below us.

That strike taught me to stay as high as I can and ensure I should approach from the coast, don't descend down the valley, wires don't tend to be over the sea and always be vigilant. It also taught me about distractions in the cockpit. We had four passengers in the cockpit — myself, a mate, my brother and his brother in law. With people in the cockpit, it is easy to get distracted and talk and become less focused on the job at hand. Recently a young pilot asked me if he could have a mate who was also flying to come along on a job with him. I had to say no as I needed him to be focused on the job at hand.

Flying is a serious job and it's something that a lot of people don't really realise unless they have been distracted and almost lost their aircraft or life.

Regarding current helicopter safety, I am worried about the current standard of training, I am concerned with the low time instructors instructing. I had instructors with 3000-4000 hours, and they passed on their knowledge and lessons to me. We have instructors with low time (usually around 240 hours) teaching new pilots. These low time instructors can only teach a maximum of 90% of what they know, then the percentage of what the new pilot takes in is even smaller. Then that instructor was taught by a 240 hour pilot also, and so on. What's even worse is the pay rate is terrible, they are concentrating on building hours and not being the best instructor they could be so the standard of training is abysmal and it's not getting any better.

Six months ago I got back into flying fixed wing aircraft, which I hadn't flown for over eight years. Incredibly, it was like riding a bike. You never really forget. I was definitely a little rusty on the finer points, but within a week I was humming again, and within the month I was able to complete a few type ratings for a number of our pilots. Sadly, when teaching some of the ratings, there was the odd component missing that had not been taught during their ab-initio training. It did rekindle an enjoyment of passing on my knowledge to the younger generation of pilot.

The industry as a whole needs to understand the difference between flying hours and currency, how current are pilots on IFR, night flying, etc. Which is a problem facing the airline industry with the incidents that have been happening — that is with pilots and the lack of experience or currency on aspects of flying.

Hours give a vague indication of experience but does not show capability at all. I can teach skills but I cannot teach attitude. We look to employ people with the right attitude and if the skills are lacking we can upgrade those skills through our training, but if the attitude is missing then we cannot do much at all.

I find Keith's comments in line with similar thoughts that has been reiterated by the other pilots. Quality training is required and the lessons learned need to be handed down from pilot to pilot.

Helicopters and fixed wing aircraft differ in form and function, but the basic principles of safe operation apply to the operators of both types. No matter what sort of aircraft you fly, everyone should be able to learn from the lessons in this book.

Finally, while writing this book I attended the HAI Heli-Expo where I saw two large drones with multiple fans and blades. One was Bell's first flying taxi model, the Bell Nexus — with six 8-foot ducted tilt rotor fans and room for five people, and the other was called a SureFly, with four arms and eight counter rotating props, seating two people.

I also visited the Hiller Aviation Museum in San Carlos (San Francisco). I always knew that Hiller Aircraft Company was part of the helicopter history, but never had an idea of how many inventions that the company actually penned and produced, in 1953 the U.S. Navy's Office of Naval Research (ONR), acting as agents for the Army, awarded Hiller Helicopters a contract for the development of a twin-engined ducted fan VTOL research vehicle named the 'flying platform'. Hiller also built and designed a no tail rotor helicopter and the flying car with four ducted fans all designed in the 1950s.

Amazing to see these concepts nearly seventy years later becoming a reality.

What lessons will be learned from these new flying machines?

I assume we will be talking about how the artificial intelligence (AI) is working and the impact it is having, maybe there won't any more pilot error? I look forward to seeing how and what limitations (if any) will be placed upon these exciting new aircraft.

And the impact these lessons have had on my flying habits? I wanted to show how powerful these stories have been for me. Many of the stories made me think about how I would approach the same situation confronting the pilot. The seat lesson is just one of the dozens and dozens of stories that made me think further about what I do, or what I need to do, to mitigate the risk before or during my flights.

I hope that when you read these stories, you'll be able to remove a few holes from the layers of risk in your flying, and build new habits to ensure that your flying is even safer than it is now.

Enjoy, be safe and blue skies.

Fletcher McKenzie

Keith's Hughes 500D is pictured on the front cover. I took the photo at the 2007 Royal New Zealand Air Force Open Day at RNZAF Base Ohakea while filming for *FlightPathTV*.

HOW TO USE THIS BOOK

A glossary of terms is included at the end of this book for your reference. Please note that this book may contain a mixture of both American English and British English, depending on who is telling the story.

If you find a term or an acronym in this book which isn't in the glossary, please email Fletcher:

fletch@avgasgroup.com

Each lesson has space for you to make your own notes if you want to. I recommend doing this to cement the learning.

Writing a short review of this book on Amazon, BookBub, Goodreads, or on your personal blog or Facebook page, will help spread the word about aviation safety. Saving lives is the primary goal of this book.

AUSTRALIA - AUS - CASA

Flight Safety Australia:
Civil Aviation Safety Authority

CASA's flagship aviation safety magazine. Topical, technical, but reader-friendly, articles cover all the key aviation safety issues – safety management systems, maintenance, runway safety, human factors, airspace, training, aviation medicine – and more.

Flight Safety Australia, and its predecessor the Aviation Safety Digest, have provided the Australian aviation community with credible and comprehensive aviation safety information since the early 1950s.

From its beginnings as a printed monochrome booklet published only a few times a year, Flight Safety Australia has evolved into an interactive and content-rich publication available across multiple digital platforms.

The website and app keeps readers updated daily. Readers can also experience a stunning interactive digital magazine version for Android and iOS tablets, available by downloading the Flight Safety

Australia app from the relevant app stores. The magazine app is published bi-monthly.

Flight Safety Australia is produced by a small, dynamic team of writers, designers and contributors based out of the Safety Promotion branch of Australia's Civil Aviation Safety Authority. You can access previous issues of Flight Safety Australia online. For editions from 1996 through to April 2014.

Close Calls. The aviation community who have had a close call write to Close Calls about an aviation incident or accident that they have been involved in (as long as it's not the subject of a current official investigation). Written by CASA Staff Writers unless noted.

With permission, we selected a number of stories for this book.

www.flightsafetyaustralia.com

UNITED KINGDOM - UK - CHIRP

Confidential Human Factors Incident
Report Programme for Aviation

Known by the acronym CHIRP, its aim is to contribute to the enhancement of flight safety in the UK commercial and general aviation industries, by providing a totally independent confidential reporting system for all individuals employed in or associated with the industries.

The Programme is available to engineers and technical staff involved with the design and manufacturing processes, flight crew members, cabin crew members, air traffic controllers, licensed engineers and maintenance/engineering personnel and individual aircraft owners/operators.

CHIRP complements the CAA Mandatory Occurrence Reporting system and other formal reporting systems operated by many UK organisations, by providing a means by which individuals are able to raise safety-related issues of concern without being identified to their peer group, management, or the Regulatory Authority.

CHIRP is a totally independent programme for the collection of confidential safety data, and when appropriate, acting or advising on information gained through confidential reports. Independent advice is provided on aeromedical and Human Factors aspects of reports, involving such topics as errors, fatigue, poor ergonomics, management pressures, deficiencies in communication or team performance. The sensitivity of these topics requires that the anonymity of the reporter must be, and always has been, fully protected.

The CHIRP organisation is comprised of a small team of specialists with professional and technical expertise in commercial aviation and Human Factors. The Programmes are also able to draw on the assistance of a wide range of individual experts and specialist bodies across the spectrum of aviation and maritime sciences in order to promote the resolution of issues raised.

CHIRP® reports are published as a contribution to safety in the aviation industry. FEEDBACK is published quarterly and is circulated in several GA publications throughout the UK.

With permission, we selected a number of stories for this book.

www.chirp.co.uk

UNITED STATES OF AMERICA - USA - ASRS

Aviation Safety Reporting System

ASRS collects voluntarily submitted aviation safety incident/situation reports from pilots, controllers, and others. It then analyses, and responds to the voluntarily submitted aviation safety incident reports in order to lessen the likelihood of aviation accidents.

ASRS acts on the information these reports contain. It identifies system deficiencies, and issues alerting messages to persons in a position to correct them. It educates through its newsletter CALLBACK, its journal ASRS Directline and through its research studies. Its database is a public repository which serves the FAA and NASA's needs and those of other organisations world-wide which are engaged in research and the promotion of safe flight.

ASRS data are used to identify deficiencies and discrepancies in the National Aviation System (NAS) so that these can be remedied by appropriate authorities. Support policy formulation and planning for, and improvements to, the NAS. Strengthen the foundation of aviation human factors safety research. This is particularly important since it is generally conceded that over two-thirds of all aviation accidents and incidents have their roots in human performance errors.

ASRS's award winning publication CALLBACK is a monthly safety newsletter, which includes de-identified ASRS report excerpts with supporting commentary in a popular "lessons learned" format. In addition, CALLBACK may contain features on ASRS research studies and related aviation safety information. Editorial use and reproduction of CALLBACK articles is encouraged. ASRS appreciates any appropriate attribution of this information. ASRS thanks the aviation community for its interest in and support of CALLBACK.

With permission, we selected a number of stories for this book.

www.asrs.arc.nasa.gov

VUICHARD RECOVERY AVIATION SAFETY FOUNDATION - VRASF

Claude Vuichard, a former line pilot and flight instructor developed the Vuichard Recovery Technique. Receiver of the Salute to Excellence BLR Aerospace Safety Award - HAI 2018.

The Vuichard Recovery Technique trains helicopter pilots to recover from a vortex ring state with minimal loss of altitude. In the vortex ring state, helicopters lose the ability to maintain lift and begin to drop.

For more than 30 years, pilots have used Vuichard's method to adjust their controls and exit the ring state by moving the aircraft to the side.

The Foundation aims at the most extensive, national and international propagation of the Vuichard Recovery Technique as well as the propagation and implementation of all further procedures and technical solutions, that entirely serve the safety of civil aviation.

The Foundation targets a purpose that serves the public, is active in a non-profitable way and does not pursue any financial gains. The Foundation acts all over Switzerland as well as abroad.

www.vrasf.org

CHAPTER 1

SITUATIONAL AWARENESS, COMPLACENCY & FATIGUE

"If an aircraft accident occurs anytime, anywhere in the world, there is an 80 percent chance that, in the final analysis, it will be due to human factors."
Dr. Robert B. Lee

LANDING FOR FUEL
CALLBACK

MD-500MG, Commercial, Phase: Cruise, Name Withheld, Apr 2012

At the beginning of the shift I performed the pre-flight inspection of the aircraft and noted that the fuel level was down below 300 LBS and was going to need to be topped off. Normally when an aircraft is left down fuel the crew leaving it down is supposed to leave a placard on the instrument panel, alerting other crews of the fuel situation. In this case there was no placard left on the panel. Due to the hot weather, I performed the pre-flight inspection in my civilian clothes (shorts and T-shirt). Once I pushed the aircraft out of the hangar, I decided it would be better to fuel the aircraft once I had my flight suit and boots on, in case fuel was spilled while fuelling. I left the aircraft without a placard on the panel and went inside to change my clothes.

Once I changed into my flight suit, I was distracted by administrative details inside the office and forgot to go back out and fuel the aircraft. Approximately one hour later, we were requested for a mission so we took off and I failed to note the reduced fuel state. I then mistakenly believed I had a full tank of fuel and I was only planning on flying for one hour. A full tank of fuel will normally allow two hours of flight time, with a reserve. About one hour into the flight I noticed a yellow caution light on the instrument panel briefly flicker. I pressed the "Test" button and realized the light I saw flicker was the "Low Fuel" caution light. I looked down at the fuel gauge and noticed that the fuel gauge was indicating below 100 pounds of fuel. I immediately realized that I had forgotten to fuel the aircraft prior to departure. I immediately turned the aircraft toward the airport and reduced power. The caution light stayed off for a few seconds and then it flickered again. I decided to make a precautionary landing in a suitable area, rather than risk having a flame out of the engine trying to make it back to the airport. Just before landing, the low fuel light came on and remained illuminated. After landing and a normal engine shutdown, we retrieved 15 gallons of fuel from our fuel facility and fuelled the aircraft. I checked and noted that we had approximately 150 LBS of fuel, which was more than adequate to safely make it back to the airport.

I flew the aircraft back to the airport without incident. I topped off the fuel tank at the hangar and the aircraft took a total of 45.2 gallons of fuel.

Upon noting the low fuel state during the pre-flight, I should have immediately put fuel in the aircraft, so it was mission ready. The reason I did not put fuel in the aircraft was because I felt I wasn't wearing the proper clothing to safely refuel the aircraft.

Lessons Learned:

To alleviate this situation in the future, I should always pre-flight the aircraft in the proper clothing, so if the aircraft needs to be serviced I will be able to take care of the situation immediately. Another option is I could have left a placard on the panel, upon noticing the low fuel state, which would have reminded me, prior to take off that the fuel level was low.

It was the first time in a while that I have flown this particular aircraft. This aircraft has a different panel layout than all of our other aircraft. The fuel gauge is a different type and in a completely different location on the panel on this aircraft. I believe this contributed to the fact that I wasn't picking-up the fuel gauge in my scan of the instrument panel.

I must make a conscious effort to adjust my instrument scan when flying this aircraft, to make sure I am not forgetting to monitor the aircraft fuel level.

NOTES:

R44 VS WIND MACHINE
CALLBACK

R44, Utility, Phase: Low Work, Name Withheld, Jun 2016

I was drying cherry trees on an orchard which I had already dried two times before on the same day. There were five wind machines in the orchard. The first two times I was drying the orchard at a speed of 10 knots groundspeed. When I did it the third time the owner of the orchard asked me to do it at 6 knots this time. When I was passing the second wind machine I must have been too close to the mast of the wind machine. I was already past the mast when I heard two hits (metal on metal). I didn't feel any hit, yaw, or change in power setting. But the noise of the main rotor blades changed to a light whistle and felt some very light vibrations from the main rotor system. Made a precautionary landing right next to the orchard, shut down and inspected the blades for damage. The blade-tip-caps were missing and there was some slight damage to the first half inch of the leading edge of both main rotor blades.

Turns out the tips of the rotor blades hit the ladder which is mounted on the mast of the wind machine. I was only one inch too close to that ladder. The helicopter had to be put on a trailer and was shipped to the next repair center for inspection.

The problem was most likely that I was flying at 10 knots on the first two times I dried the orchard that day and then had to slow down to 6 knots on the third time. That might have made me think I'm already past that mast and I came back to the tree row a little bit too early.

Another thing was that I was worried about, if I had enough fuel in case the owner wanted me to do another pass over his orchard after I did it already three times. That might have distracted me from my task to fly safely over the trees and stay far enough away from the obstructions.

Lessons Learned:

Next time I have to focus on keeping my distance from all obstacles (even though it seems like I've already passed them) and don't let anything distract me from my job. Worry about the fuel calculation when on the ground.

NOTES:

TAKING OFF STILL ATTACHED
CALLBACK

Unknown, Emergency, Phase: Take-Off, Name withheld, Jan 2001, Issue 257

I was going to move the aircraft to airport so it could be hangared from the approaching severe thunderstorm. The aircraft has several orange electrical cords used to power the medical equipment and cellular telephone.

I walked around the aircraft, untied the rotor blade and observed the orange cords lying on the ground. I started the aircraft and took off for the airport and returned to the hospital due to the thunderstorm.

I missed one of the cords plugged into the aircraft and it became tangled in power lines on approach (over) to the hospital. No damage to the aircraft occurred.

Lessons Learned:

The incident was caused by the urgency to move the aircraft due to severe weather... I also started the aircraft without the assistance of the other crew members, as they were busy. All of these factors caused me to miss the cord going in the right rear door of the aircraft.

<u>NOTES:</u>

DOG DAY AFTERNOON
CASA

R22, Private, Phase: Landing, Craig McIntosh, Apr 2008

I am a private helicopter pilot and have accumulated about 340 hours over ten years. Most of my flying is done in and around the Southern Highlands of NSW, where I live. My flying set-up is close to perfect and fits into a neat routine: I rent an R22 from Bankstown Airport and fly it down to Mittagong Airfield, where I keep it in a friend's hangar. From there I do outings alone, with my wife, or with friends.

I have a lovely old diesel Landcruiser that is my transport to and from the airfield, and I am always accompanied by my two kelpie dogs – Impi and Jasmine – who love the airfield and run amok whilst I pre-flight, refuel or wait around for weather conditions to improve. The three of us are a well-oiled flying team and have great fun together!

To brush up on my aviation knowledge, I have recently done three exams towards my commercial licence, including 'Human Performance Limitations'. Theoretically, I am aware of the many issues surrounding the human mind and the importance of pilot judgement and temperament. Since writing that exam, all of my flying had been uneventful on that front. That was about to change.

On a recent, beautiful spring day, I got the itch in the late afternoon to go flying. I wanted to revise a few skills and to improve my concentration. The dogs came along, with Impi hanging out of the passenger window as we wound our way along the country roads to the airfield.

After the pre-flight, I left the dogs in the cabin of my ute with windows wide open for fresh air, started up the R22 and flew away. What bliss! It was a perfect early evening with cool, calm air and a stunning sunset. Because of this, I flew on a bit longer than I had first planned, soaking up the joys of nature as well and the wonder of powered flight! After a good 30 minutes I turned for home, opted for a straight-in approach and set myself up on long final for Runway 06. As I did my final checks, I could see that the airfield was deserted except for my ute which was parked in front of the hangars.

Once in the hover, I decided to make the most of the fading light and do a few drills – hover turns and 'square bashing'. As I came out of the first left-pedal turn, I noticed a rapid movement out of the corner of my eye, and as I turned out of the setting sun, realised much to my surprise that my kelpie dog Impi was out of the ute and heading over to see his 'dad'! I was not that concerned, initially. I couldn't imagine that he would come close to such a noisy machine, and if he did, I would simply hover away from him and set down elsewhere.

Within seconds, he was at the tip of my right skid (well within the wash and noise of the main rotor), with his eyes locked on the love of his life – me. It was also obvious that he wanted to hop in for a ride. My main concern now was to keep the tail rotor away from him. I backed away and darted across to the hangar apron and turned to face where I had come from – with the setting sun in my eyes again. Impi is a fast dog, and he was already at my right skid. Now what? I hovered much further away this time. Same outcome! And during all of this, I was yelling as loudly as I could at him to get away, even opening the air vent on my door to help carry my voice to him. But that was all to no avail. The engine was drowning out my voice and I could feel my throat burning from the effort.

The next idea that popped into my head was to do a circuit and give Impi time to get away from the landing spot, and hopefully jump back into the ute. The sun had set, so light was fading fast. On final approach, I could not see him anywhere. But as a precaution, I went well beyond the hangar to the refuelling shed.

As I turned in the hover to check my luck, there he was – at my right skid! I was now getting very annoyed. Next plan? I decided to actually chase him with the helicopter, but that did nothing except raise the ire of two nesting plovers, which now joined the fray by circling wildly around me and Impi, mainly trying to dive-bomb him, but at the same time coming very close to me and distracting me. I was not so keen on a plover hitting the tail rotor or the main rotor, so I now found myself in a high hover watching two swirling birds, one dog, and darkness settling in.

I had to do something, and the right solution finally came to me. I broke away from the commotion that I had created and flew as fast as I could to the far end of the runway, landed as quickly as I could, turned off the governor, locked down the cyclic and collective, then looked up just in time to be able to open my door, grab Impi by the scruff of his neck, heave him up into the helicopter with me, and slam him down into the passenger foot-well!

Once inside the machine, Impi was suddenly terrified and did not move an inch. I was therefore able to complete my shutdown as per the checklist, and lock up the helicopter for the night.

It was totally dark by now, I was too far from the hangar to wheel the helicopter under cover (as I should do), and my throat was raw from yelling. I was very worked up and irritated, and Impi was trembling like a leaf. The good news is that he has not gone near the helicopter since – even when the engine is off and it is casting the only available shade on a hot summer's day.

Lessons Learned:

I look back on this incident and think that the situation was not that serious; my take on the whole experience is how quickly situational and human performance factors can come into play, when you least expect them – but when they are most needed.

Within the space of a few minutes, I had gone from enjoying a relaxed sunset flight to being caught up in a kelpie and plover obstacle course which, for a while, left me totally flustered and without solutions. At the end of it all, I was also caught out in the dark, and when driving home, realised that I had not looked for other traffic (albeit unlikely) as I rushed to the far end of the runway to shut down.

I guess I was lucky that the situation was perhaps more unexpected than dangerous, and that Impi was actually brave enough to come right up to me where I could grab him and put an end to the Kelpie caper. On the night of the incident, Impi was very much in the doghouse. But perhaps it should have been me?

NOTES:

UNPLANNED TOUR STOP
CALLBACK

R44, Private, Phase: Cruise, Name Withheld, Jan 2012

Customers were being loaded and unloaded without engine shutdown in order to save time and enable us to complete the tours within their time and budgetary constraints. Tours were also taking longer than anticipated... As the last group was being loaded, I should have done a more complete assessment of fuel quantity and reserves and shut down to obtain more fuel... The low fuel light illuminated approximately eight miles from the airport... I found a flat place and set down to call for fuel.

I over-emphasized weight considerations, not filling the tanks completely before the tour group arrival in case the passengers turned out to be heavy. Extensive experience in the R22, for which five gallons is an adequate fuel reserve, and less cross country/long flight experience in the R44, for which that reserve is inadequate, made me tend toward an insufficient estimate of what I needed. The wish to keep the tours moving along and save my customers time and money made me rush both moving forward with the first tour (when I could have stopped to fill the tanks after realizing that we were not weight limited) and also made me reluctant to stop in the middle of the series of tours.

Lessons Learned:

Unexpected tasks in the office that morning filled up much of the two hours of preparation time I had allotted myself to prepare for the flight.

Rushing through the preflight phase made me more likely to miss the error I had made in the fuel calculation.

<u>NOTES:</u>

CONTRARY CONTROLS
CALLBACK

BK-117, Commercial, Phase: Pre-Flight, Name Withheld, Jan 2012

I arrived at work for the night shift, drove to the helipad and performed what I thought to be a thorough preflight. It was dark and it was misting due to an approaching thunderstorm. I used a flashlight and I paid particular attention to the maintenance that had been done that day involving an engine fuel pump replacement.

I was aware that the dual controls had also been installed that day per the logbook entry and that both the fuel pump and dual control installation had been checked by the day shift pilot.

I completed the preflight noting nothing out of the ordinary. I did not fly during my shift as the weather was below my minimums.

After leaving work the following morning, I drove home and later that day received a call from the day shift pilot informing me that a mechanic had installed the left cyclic backwards and that we had all missed it.

The day pilot found the error prior to any damage or injuries while preparing to depart on a patient transfer flight.

Lessons Learned:

Suggestions:

1. Each pilot should sit at the positions that have a set of controls during preflight to ensure that the controls are mounted correctly and a full flight control function check (at each station) should be performed each time the controls are removed and reinstalled.

2. To completely avoid this error it may be appropriate for the manufacturer to re-engineer the way flight controls are mounted so that they can only be installed the correct way. It could have been a serious problem had this error not been detected prior to takeoff. If the left seat was moved forward, movement of the cyclic could be hindered by contact with the seat.

<u>NOTES:</u>

ROTOR RAGE
CASA

500D, Fire-Fighting, Phase: Load pick up, Name withheld, Apr 2006

The spring of 1980 in north-west Ontario had all the signs of a major fire season. In May, a heatwave settled over the broad expanse of lakes and forests and each evening towering cumulus would rapidly build, often flattening out to an anvil-head at the 40,000 ft level. More often than not, lightning would flash into the evening with no rain.

A friend of mine had built up his small helicopter company from one Bell 47 to two Hughes 500C models, and he had lured me away from Canada's largest operator of Hughes aircraft on the promise of becoming chief pilot. Plus, he needed someone with 500D time as he'd just acquired his first, brand-new helicopter.

With 21 hours on the airframe, I flew this brilliant Hughes 500D C-GGSX from Winnipeg to Dryden, Ontario, and waited to be dispatched. Dryden was the regional headquarters for the Forest Protection Branch of the Ontario Ministry of Natural Resources. From here all fire-fighting resources – ground crews, equipment, water bombers and helicopters – were allocated to various fires.

On the afternoon of the very next day a forest fire (Kenora #23) erupted, and with the fire-boss on board we departed. This being the maiden flight of my friend's 500D, I was keen to show it off and we zipped along just under VNE. It was a tight helicopter, clean and humming with power. Because we had just taken possession, our engineer had not had time to add some basic kits to it – a cargo mirror on the toe of the left skid, and bear-paws to keep the tail up when landing on soft ground.

The position of the fire on the map (we had no GPS then) put it just south of the Trans-Canada highway. About 20 nm east of that point the smoke forced me lower and slower along the road, until I had to break away to the south. This was no spot fire – it had begun to spread rapidly to the north-east.

The fire-boss passed along information to Dryden, who then mobilised crews and equipment. We scouted out a base-camp location, guided ground crews into the initial attack areas and mapped the fire, which seemed to be growing exponentially. Fuel trucks and crews arrived by the minute and were dispatched to various locations. Helipads had to be cut, homeliness laid, fireguards bulldozed.

No regard: I flew until dark that first day, excited and thrilled to be the first helicopter on site. My friend and employer would be pleased. When I asked where I would be sleeping that night – we were 20nm from the nearest town – the fire-boss shrugged and said, "Anywhere you can find a spot."

Our base camp was a collection of old huts, but fire crews and administration people, who had arrived while I was flying, had taken all the cabins with beds in them. The main building provided office space on the ground floor for the command-post personnel, with empty rooms upstairs designated as sleeping quarters. Cots were on order.

I found some stale white bread and a slice of processed cheese in a fridge for my dinner that first night. Luckily I had brought my own sleeping bag, so I curled up on the floor of one of the empty rooms upstairs. Fire fighting personnel arrived at all hours, and judging by their raucous greetings and thumping as they entered the command post, they considered sleep a non-essential luxury.

At dawn the next day, after a quick and distasteful instant coffee, we were airborne once again for another survey of the fire. It had grown during the night and more resources were called in. Meals had been brought in for the command post staff but apparently not enough for a hungry pilot. Again I had to scrounge around for food. Cots had been delivered and set up in the rooms upstairs – eight cots to a room – so I had a bed to sleep in that night, but again, senior staff arrived at all hours and stumbled into the rooms without much regard for those already asleep.

Around day three of the fire, a mess hall had been established and occasionally I had time to sit down for a greasy pork chop, some fries and soggy peas from a can. I complained about not having scheduled meal times and a proper place to sleep, hoping they would release me each evening to fly to town to stay in a hotel. Many of the medium and heavy helicopter pilots were in town, and of course, all the fixed wing bomber pilots.

But I was "essential crew" said the fire boss – his personal air taxi and he wanted me close at hand. Eventually I had a room with only four cots in it, and time for breakfast and dinner. Lunch was often eaten in the air as the afternoon heat usually created havoc out on the fire lines.

Backfire: On day six we went out and set a backfire – a procedure that should have stopped the spread of the fire. Of course, the wind shifted just after we ignited an area of 800 ha and within two hours it had erupted into a raging inferno of another 16,000 ha. Airliners had to deviate from the column of smoke. The intense heat caught many fire fighters out of position and much of my day was spent pulling crews out of intense and potentially dangerous situations. Many miles of hose and numerous pumps were burnt up – but we got the men out safely.

This only fed my enthusiasm for the job, and a 10-hour flying day passed almost unnoticed. The next day I logged 9.5 hours as government clerks and accountants were pressed into service as emergency fire fighters. Men were being pulled from the bars, vehicles were stopped along the highway and all able-bodied men were conscripted to take over unskilled jobs so employees with "some" training in fire fighting could be better utilised.

Day eight on Kenora fire #23 was another clear day with gentle winds from the SSW at 12 kt. The fire-boss had scheduled a day of rest for me – after a simple crew move and one net-load of gear. Somewhat disappointed that I was going to miss out on another lucrative day, I nevertheless appreciated that I was tired and needed a break.

At 07:45 I departed for a helipad 10 minutes away with a crew of five on board. I dropped them on the edge of a small lake and told them that I would return in about 20 minutes with their gear.

A large net had been loaded in the centre of a grassy field. One fire fighter in regulation-issue orange coveralls stood near the net holding the short 3-foot lanyard. About 20 m away another orange-suited fire fighter had positioned himself to act as marshaller. All was well in the world.

I lowered the Hughes 500D over the net and stabilised in a low hover, expecting a quick and efficient turnaround. I sat in the hover for what seemed an eternity and still had not received the thumbs up from the marshaller. I had no mirror so I couldn't see beneath me.

The marshaller then directed me forward and to the left. He wanted me to go higher, then lower. He moved me to the right and back a few feet. I found this rather odd because I knew I had put the hook directly above the net on my first approach. Come on, I said to myself. The crew is waiting for this gear.

Again, the marshaller directed me back, forward, left and right. The guy beneath me only had to pull on the lanyard and attach it to the hook. What was the problem? I thought perhaps the marshaller needed glasses.

Growing more frustrated by the minute, I signalled to the marshaller that this was no good; I was going to do a circuit and come around again. I shook my head in exaggerated movements, took my hand off the cyclic and made a quick circling motion with my finger to indicate a circuit.

I remember thinking: "I'll show you." And in an act of sublime idiocy – like road rage – I jerked up on the collective and shoved the nose forward. The 500 leapt into the air. I was going to muscle the sporty machine around in a tight circuit and put the cargo hook right over the net, again.

But as the disk tilted forward, the helicopter pitched up violently, lurching to the left. All I could see was a brilliant blue morning sky and a blazing sun. A great silent and pleasant calm washed over me, as though some major decision had been made. Aware that someone was below me, I pulled more collective to suck the aircraft away from him.

The impact was a blur, and then everything moved in slow motion. Strapped in upside down, the windscreen shattered, I saw the illuminated caution lights and heard a beeping. I found the collective underneath my left leg, covered with dirt, and turned the throttle off. I reached forward to turn off the battery. It seemed like minutes had gone by.

Suddenly, the fire fighters were reaching in for me. They sounded panicked, yelling that the helicopter was going to blow up. I tried to reassure them it was fine.

I released myself from the shoulder harness and crumpled to the ground. They pulled me out and I hobbled away with their help. I had bruised my left leg, but other than that I was fine – physically.

I found out a few minutes later that the two "fire-fighters" assigned to hook up the load were not fully trained. I had assumed they were, with dire consequences. With no mirror and no bear-paws, I had no way of knowing the net had become entangled on the back of my left skid. My impatience and rapid increase on the collective set in motion an unrecoverable dynamic rollover. And that was the end of that brand new 500D.

Of course, the long chain of events leading to the accident taught everyone some valuable lessons. Some things have improved since then, and others have stayed the same. For myself, I have since enjoyed 25 accident-free years, mastering the slow, methodical lift on the collective.

Lessons Learned:

CASA Comment: This accident took place some 26 years ago and I would hope that many of the practices that were common in those days would not occur in Australia or other parts of the world today. Even so, there are a number of relevant lessons here, both for helicopter pilots and the aviation community at large.

Fatigue:

First and foremost, the story is a reminder of the insidious effects of fatigue. In this case inadequate rest facilities, poor sustenance and an arduous duty cycle created a situation that adversely affected the pilot's performance. Common symptoms of fatigue include lack of concentration, impatience, a preparedness to cut corners and settle for lower standards and poor decision-making. All of these were seen in this accident, finally culminating in the pilot's "dummy spit" in which he hurriedly increased collective and unwittingly instigated an unrecoverable dynamic rollover. Fortunately, no lives were lost and the pilot lived to tell the tale.

The current emphasis on accountability and human factors in the fire-management industry acknowledges that fatigue is a major hazard that must be addressed. Compliance with the Part 11 exemption to CAO 48 is one way to avoid the kind of fatigue experienced by the writer. Similarly a company fatigue risk management system (FRMS) is an effective means of identifying fatigue risks specific to the company's operations so that practices and procedures can be put in place to defend against them.

Training:

Would the accident have occurred if the ground crew were properly trained ? It's the responsibility of the operator to ensure that ground personnel are properly qualified and briefed. The briefing should reinforce training already provided by the fire management authority, and include a discussion of safety issues around the helicopter and clarification of correct hand signals. The operator should never assume the training has already been provided.

Equipment:

If the aircraft had been fitted with proper mirrors, it's probable that the pilot would have seen that the net was hooked on the skid and would not have reacted as he did. Further, if the ground crew had been equipped with a VHF radio, the problem could have been easily communicated to the pilot. The pilot would then have been in a position to direct the ground crew to fix the problem. Another factor was the decision to use of a 3 ft sling to attach a cargo net to a Hughes 500. A 10 ft sling (or longer) would have allowed a hook-up well clear of the load and eliminated the risk of snagging a net.

There is no excuse for commencing a task without adequate support equipment and preparation; you do wonder though if the pilot would have accepted the same risks if he had been well fed and rested.

Mal Walker
CASA Flying Operations Inspector

<u>NOTES:</u>

REST IN PIECES
CASA

Bell 206, Commercial, Phase: Load drop, Name withheld, Jan 2004

I don't know how long I lay there with my face in a pool of Jet A1 fuel, and an upside-down Bell Jet Ranger sitting on top of me. I was aware of a whistling noise from somewhere behind me, and a fellow asking me if I could move.

I was the manager of a remote helicopter base. We had started out with two aircraft, three pilots, an engineer, an apprentice and an office manager. We had a mixed client base including a port authority, several mining companies, and a number of small survey organisations. There was plenty of competition for the available work, and margins were tight. The first casualties of economic rationalism were the office manager and the apprentice, followed shortly by the third pilot and the engineer.

The work load reduced to a degree but we were still on 24-hour standby for the port authority. It's difficult to run such an operation with two pilots, but I was assured the authority (then the Department of Civil Aviation) knew about it and thought it was OK.

To complicate matters, our commercial manager had stressed that we were not to reject any potential jobs – after all, money was tight. I suppose I should have closed the office, left the stocktaking to someone else, delegated more duties to the second pilot, given the base car to someone else to fix and taken more time off, but I was trying to give the impression that it was business as usual and that I was coping.

A regular client had a sling load of radio equipment to go to an island off the coast and I tasked the second pilot for the job, making sure that he carried the EPIRB and had the chin mirror fitted. A week later our client called to request a pickup the following day, which was a bit of a nuisance as it was the second pilot's day off and I was looking forward to a free day myself – but I couldn't knock the job back.

I got to the island and picked up the sling load but realised that I had forgotten to fit the chin mirror. On the return trip, the EPIRB (which I had remembered) fell off the hat rack and on to the floor – I'd neglected to secure it. That was two strikes.

When I returned to the airfield, I just wanted the trip to end. I didn't notice the marshaller who was there to direct me, so I placed the load on the ground and moved to the right of it to prevent the clevis from damaging the radio equipment when it was released from the hook. I pressed the hook release and kept moving to the right while looking underneath the helicopter for the released clevis (a mirror would have been nice).

Strike three: the clevis wasn't there, the hook had failed to release. That was when a bad day turned really bad. The sling, which was attached to the load, was about two metres long. When I moved right it became taut and jammed in the throat of the hook, keeping it closed while the sling lifted the left skid. The hook remained fixed in space while the helicopter proceeded to enter a classic dynamic roll over. There was nothing I could do but accept my fate .

When the noise stopped, all I could think of was how nice it was to have a rest. In hindsight , it is easy to see what happened: I didn't question my superiors, I didn't obey the law, and I didn't get enough sleep. I was complacent and even overconfident. I thought I could handle fatigue, but it beat me. Fatigue is insidious. It crept up and clouded my judgement without my knowing. There were signs, but I didn't recognise them.

Lessons Learned:

That's why the rules are there: to protect us from ourselves and from those who would take advantage of us. I learned a lot about flying from that accident. Here are some pointers that will hopefully stop any of your days from turning really bad:

- Understand the causes and effects of fatigue.
- Plan ahead (even the simplest activities may have elements of risk that can be reduced).
- Know your equipment and its limitations.
- Don't be afraid to seek advice and assistance from others – mortals cannot do and think of everything.

I was incredibly lucky to survive. In fact, I returned to flying within a week of the accident. I lost an aircraft, a client and an incredible amount of self-esteem, but I will never take safety for granted again.

Analysis: Don't fly tired.

CASA Comment: Fatigue has long been recognised as a serious threat to aviation safety. Since the 1950s, Australian aviation law has prescribed maximum flight duty times and minimum rest periods, with the aim of protecting pilots and their passengers from the effects of fatigue. These restrictions are detailed in Civil Aviation Order 48. Variations to CAO 48, in the form of exemptions, often provide for even greater rest periods when arduous duty periods are expected or the potential for sleep disruption exists.

Fatigue risk management systems (FRMS) are gaining acceptance as an effective way of tackling fatigue. FRMS take a holistic approach and look at, among other things:

- hazard identification and mitigation
- education
- fatigue measurement
- rostering
- management commitment to FRMS.

Fatigue risk management acknowledges that, despite our best efforts, fatigue can never be completely eliminated. Therefore, organisations need to establish defences to reduce the consequences of fatigue. Typical defences might include extra checklist requirements, more-stringent procedures or a requirement for additional crew members during high-fatigue periods, such as in night operations.

In 2001, CASA began a trial of fatigue risk management systems with 16 general aviation organisations comprising fixed-wing, helicopter and balloon operators. The University of South Australia's Centre for Sleep Research evaluated the trial and found that, "Approximately 90 per cent of managers, and 85 per cent of flight crew members perceived the FRMS had a positive impact on operations". Although most operators recognised the benefits of FRMS, including increased flexibility and safety, there was concern about difficulties in implementing the systems and "significant upfront costs".

No system is perfect and it's important that pilots, whether they operate under CAO 48 or an approved FRMS, use prescribed rest periods to obtain adequate sleep. The only way to reverse the effects of fatigue is to sleep. Learn to identify the signs of fatigue, and, above all, do not fly if you are tired.

Senior managers must recognise the dangers of stretching resources and adopt appropriate strategies to minimise and defend against fatigue. Whether it was intended or not, the management of this company created a culture in which the pilot thought it would be better to fly exhausted than to reject the flight. In the end, the commercial manager's instruction "not to reject any potential jobs" turned out to be very costly.

Mal Walker
Flying Operations Inspector

For more information, visit the CASA website at www.casa.gov.au/avreg/ business/fatigue

NOTES:

A WRINKLED TAIL BOOM
CALLBACK

AS350, Training, Phase: Landing, Name Withheld, June 2010

This flight was a training flight with one of our new hires. There were no issues with this student and their flying skills. We were going to be covering basic manoeuvres and some emergency procedures.

After briefing the student with the lesson plan and illustrating what to expect on the flight, we proceeded out to the ramp for the flight. The flight was conducted under VFR flight rules, and lasted a total of 1.4 hours.

We departed for an outlying to conduct the flight due to the amount of traffic at our home airport. We joined the outlying field traffic pattern and started off with basic manoeuvres, normal approach, air taxi, quick stop, max P take-off, and steep approach. Then I started conducting some emergency procedures such as stuck power pedal, and hydraulic failure.

These two procedures require a shallow approach running landing. I conducted two hydraulic failure procedures because I felt that his first approach for this procedure was not shallow enough. After the first running landing for the hydraulic failure and we were stopped on the runway, I turned the hydraulics back on. I then explained to the student to shallow out his approach to make the emergency landing easier. I then took controls, picked up to a hover and back taxied closer to the approach end of the runway. We then moved on to some slope landings in the grass on the west side of Runway XX.

After the slope landings, we moved on to some hover auto rotations over grass as well. This is where I think the damage may have happened. Before conducting the hover auto rotations, I had the student set down the aircraft so that I could confirm a level surface.

We conducted 5 hover auto rotations over this spot from what looked to me a 3-5 foot hover height. The first one was descent, but the student was not working the pedals to maintain heading. The 2nd hover auto, we started to drift forward and to the left slightly and we did not land as smooth as I would have liked it. I then explained to try to maintain current position over the ground and just to let the aircraft settle before pulling collective pitch to cushion the landing.

The 3rd hover auto, the student did the same thing, forward and slight drift to the left with a rougher landing than I would like, but initially controlled the heading much better when I reduced the throttle, well within commercial PTS standards. At no point in time after these hover auto rotations did the idea of a hard landing go through my mind. Everything felt normal to me in the cockpit.

The 4th and 5th hover auto, I pretty much demonstrated the manoeuvre myself with a very smooth and gentle set down. At this point of the flight, the student and I were quite hot in the cockpit and proceeded to rejoin the traffic pattern. We finished off the training flight with two straight in auto rotations power recovery and three 180 degree auto rotations power recovery. Flight characteristics of the aircraft seemed to be normal.

We returned to our home base to end the flight training lesson. We parked the aircraft in front of the home base hangar and proceeded for normal shut down the aircraft. After shut down, and once the blades had stopped turning, I got out of the aircraft and proceeded into operations to tell the radio room what my landing time was because I could not talk to them on the proper frequency. I then headed back outside to get with my student and he had several questions about the flight log.

After helping the student, we grabbed our things from the aircraft and proceeded inside for a debriefing. This is where I know I made a big mistake. I did not do a post flight. If I had I would have seen the damage and taken the proper action, but this is not the case.

Unfortunately, I found out about the damage from my student when I was heading back out to the ramp to speak with my boss. I do not know who noticed the damage first, but as I was walking out to the ramp, I noticed several people around the aircraft. Once my student told me about the damage, I proceeded out to the aircraft to observe the damage. I was in disbelief about the damage and began trying to think about when this could have occurred. Once again, I did not notice any abnormal flight characteristics of the aircraft throughout the entire flight.

Lessons Learned:

Do a post flight check.

<u>NOTES:</u>

LOW FUEL LIGHT
CALLBACK

R44, Training, Phase: Landing, Name Withheld, May 2016

I was concluding a lesson, and decided to fly to the coast (5nm) when we departed I had 1/4 tank in the main. About 7 gallons. About 25 min of fuel.

We flew [to] the coast and I monitored the fuel as we did, we flew for roughly 14 nm at 100kts and then turned around. When we were entering the airport airspace, the low fuel light began to flicker, I noticed we still had 1/8 of a tank. The entire time I was judging our fuel based on the gauge. I called tower and told them I was fuel critical and requesting a direct landing at our hangar.

Tower told me to continue, and asked if I'd like to declare an emergency, I said no. About 4.5 nm from the airport the light was steadily on. I told my student to count down 5 min. A mistake I made, confusing the R22 low fuel light flight time remaining. The R44 is 10 minutes. Something I should have known by heart.

Once my 5 min were up, we were 1 mile from the airport, I selected a landing spot and began my approach. I advised tower of my intentions. The entire time I had a road on my left side, should I lose my engine. Once landing was assured, I told tower. Tower said that emergency services were on their way. I called my boss, and told him about the situation. I then told tower our specific location and told them we would turn off the master now. We called someone to bring us fuel and waited for the emergency services.

Once we were fuel up, with 10 gallons, we advised tower and joined the downwind and landed at our hangar.

Once back at the airport, I had the fuel truck top off both tanks to see how much I had left. It was 3.2 gallons. About 12 min at 15 gallons per hour. That was the amount left when I landed shy of the airport. I had a meeting with my boss two days later and we discussed the situation. We concluded that I made the right decision to land, but made a very bad decision to depart with low fuel.

Lessons Learned:

I broke a regulation by departing with not enough fuel. And I trusted my gauge too much. I will never leave with 1/4 tank again and will always land with a minimum of 1/4 tank.

<u>NOTES:</u>

COLLECTION FRICTION ADJUSTMENT

CALLBACK

R44, Training, Phase: Parked, Name Withheld, Apr 2016

I was taking a Rotorcraft CFI for his initial flight in a Robinson R-44II. This instructor had completed all of his time (+/- 180 Hours) in the Schweitzer 300. He was interested in getting familiar with Robinson helicopters to expand his employment opportunities. We completed the required SFAR 73 awareness training beforehand. Next we walked through the preflight checklist step by step taking time to become familiar with aircraft systems. Utilizing a checklist again, the engine start was uneventful, however while explaining the purpose of the hydraulic system check the engine stopped unexpectedly. We discovered quickly that when removing friction from flight controls the student mistook the main fuel shutoff for the collective friction adjustment. Leaving the collective friction on and the fuel off.

 The hydraulic system check is one of the last checks you perform prior to liftoff. If we had been in more of a hurry to depart we could possibly have been airborne in a hover or takeoff profile when the engine stopped. In my opinion this was a possible near miss for an incident and an excellent learning opportunity. Personally, I check the friction tension prior to liftoff to ensure no hindrance of flight controls. I am sure that prior to liftoff I would have checked the collective friction and noticed the fuel shutoff in the closed position. Also we possibly would have not been able to complete the hydraulic system check with the collective friction still applied.

Lessons Learned:

In the future I will be taking extra steps to ensure each student fully understands the importance of knowing the difference between the collective friction and the fuel shutoff. There are multiple checks in place to help avoid this from causing a fuel starved engine in flight and I am unaware of any similar instances but I feel that this may possibly be something that others need to know about in order to avoid a more disastrous outcome.

<u>NOTES:</u>

MOMENTARY LOSS OF CONTROL
CALLBACK

R22, Training, Phase: Cruise, Name Withheld, May 2015

I set out on a solo cross-county flight as part of my helicopter private pilot training. The aircraft was a Robinson R-22 Beta II. All of my recent flight training had been in this aircraft. My primary flight instructor reviewed my flight planning and endorsed my log book for the flight. The aircraft tanks had been filled (29 gallons).

I departed at about XA:45, 45 minutes late and 15 minutes past the departure time on the VFR Flight Plan. Once given clearance for frequency change by Tower, I contacted FSS and opened my flight plan. The flight initially followed Highway 101 and then crossed inland in a roughly straight line to Santa Barbara Airport. (On the advice of my instructor I had chosen not to follow my plotted route via VORTACs because of mountainous terrain.) Within 25 miles of Santa Barbara I contacted Approach control and was given a squawk. Within 6 miles of the airport I was handed off to tower who gave me clearance to land on the FBO helicopter pads. The approach and landing was normal but for adjusting the flight path to avoid a tall crane on the highway meridian. After setting down, I adjusted the radio frequencies and GPS NAV unit for the next leg to Camarillo Airport, as well as sending a quick text message to my wife. Departure was along the same route and I was passed to departure control. About 10mi from the SBA I was granted a frequency change. I followed the coast and Highway 101 until passing inland to arrive north of Ventura and Oxnard Airport. I began listening to Camarillo ATIS but it was spotty 20 miles out and low over the foothills.

I passed over Lake Casitas and Highway 33, moving over the last of the low hills before the flat Oxnard and Camarillo area, about 15 nm from the Camarillo airport. I was flying at approximately 800 feet AGL (1,300 feet MSL), airspeed was approximately 80-85 KIAS, Manifold Air Pressure (MAP) approximately 19-20 inches, the collective had the friction lock engaged and the cyclic trim knob was up. Time was approximately XC:15 PM. I began preparing for my arrival in Camarillo by pulling the airport diagram from the stack of papers on the left seat. With my left hand I tried to insert this under the clip of the knee board I had on my left thigh. I dropped the diagram under my right leg and could not reach it with my left hand. So, after some consideration, I grasped the collective crossbar with my left hand about half way between the column and the grip so that I could reach down with my left hand. I had never held the cyclic in this manner but, given that I was in cruise flight and felt I could hold the control with suitable firmness, I made the attempt. I had seen instructors make adjustments to my inputs in this manner. The mechanical advantage and fine control proved lacking.

In reaching down I made a nose-down input that made me jerk up and reach for the grip again with my right hand. In releasing the cross bar with my left and reaching for the grip with my right the control suddenly jerked forward and to the left out of my tenuous grasp. The helicopter snapped over to about 110 degrees left roll and the nose dropped approximately 70 degrees below the horizon.

I quickly grabbed the cyclic and centered the column before rolling upright at a moderate rate and then pulling up gently. I had also grabbed the collective lever. In scanning the instruments I found the collective had dropped to 15″ and I was crabbing a few degrees to the right, both which I corrected. All else appeared normal. I was too busy looking out at the aircraft attitude to notice the RPM. There was no unusual vibration or noise so I continued the flight. I noted that papers in the left seat, weighed down by my cell phone, had not been disturbed, suggesting I had remained in positive G. I did not feel any unusual unload.

I had difficulty finding Camarillo Airport at the low altitude I am used to flying in a helicopter. I was in contact with the tower and they advised me about traffic which I identified above me. Communications were difficult on both radios owning to intervening terrain and they lost my transponder signal a few times as well. Finally, with good comms and transponder signal, the tower vectored me to the field. I followed tower instructions to the helicopter pads (the airport diagram remaining on the floor), I set down and again prepared the com nav systems for the next leg in addition to sending a text to my wife. Two men in a golf cart waited about 50 feet to my right side to pass and I waved to them and they returned the greeting. They twice made a hand signal I could not interpret but they may have been indicating that the rotor was drooping a surprising amount at the 55 RPM setting. They were smiling and not overly insistent so I dismissed this. Eventually they drove past my nose as they realized I was not immediately going to move. Again, the helicopter's vibration was normal and there were no unusual noises. (I did not remove my headset to listen but had never done this before so would have been unfamiliar with the sound.)

I departed Camarillo following tower instructions until given permission for a frequency change. I followed the 101 into the San Fernando Valley. I initially gave erroneous position reports to Tower but eventually sorted this out for arrival. Approach was normal and company instructor, preparing an R-44 for flight with a fuel truck nearby, waved me over to taxi across an unused ramp to the parking spot. This required me to approach and hover over the spot and then make a right pedal turn to set down. Shutdown was normal without unusual noises (headset still in place). The flight was complete at XD:oo PM (30 minutes late) for a total flight time of 2.3 hours.

It was only after I stepped out of the machine that I noticed a pronounced blade droop. My wife who greeted me after shutdown, said that she heard the blades brushing against the tail boom. Examining the boom I saw only three faint witness marks on the right side below the blade tip path, but these may have been unrelated. No other damage was evident. Subsequent examination found that the teetering droop stops had been broken and that I had come very close to catastrophic mast bumping. I forgot to close my flight plan in the flurry of post-flight activity. I phoned the NTSB Los Angeles office at XE:oo PM to make a verbal report of the accident.

The last flight before the mishap had been in a fixed wing airplane the previous day. The last helicopter flight had been in the mishap aircraft four days earlier. The only prior solo flight had been three approaches three weeks earlier. Total rotorcraft time to the date of the mishap was 30.1 hours that included 1 hour dual cross country and 1 hour dual night. My total time is roughly 950 hours over 33 years of flying.

Lessons Learned:

My airplane aerobatic experience allowed me to remain reasonably calm (apart from a loud expletive) and avoid over-controlling the machine.

I had recently reviewed the flight manual verbiage about abrupt manoeuvres, especially unloaded pushovers with the potential for mast bumping, followed by abrupt pulls with the risk of boom strike. I was surprised that neither of these events had apparently occurred as the helicopter remained together and I had heard no load bang or felt anything like a strike.

NOTES:

CONTACT WITH TERRAIN

CALLBACK

R22, Training, Phase: Final Approach, Name Withheld, Mar 2003

My student and I were flying about 1,000 ft AGL. Upon selecting a pinnacle to make an approach to, we began our high orbit to conduct WOTFEEL (Wind, Obstacles, Turbulence, Forced landings, Entry, Exit, Landing). We determined the wind to be out of 300 degrees and began our approach once our checks were complete. Once on final approach at around one hundred yards from the landing zone, we encountered a small downdraft to which the student raised the collective to arrest our descent.

Though our approach angle was unaffected and rotor RPMs still at 104%, I instructed the student that his input of up collective was too great and that it needed to be smoother.

At around 100 ft from the landing zone, we encountered another small downdraft to which the student aggressively raised the collective, instantly dropping rotor and engine RPM to approximately 95%. I immediately took the controls from the student and initiated low rotor RPM recovery procedures of rolling on the throttle and applying slight down collective. Unable to recover rotor and engine RPMs, the helicopter began to descend towards the landing zone.

Despite my best efforts to remain airborne, the helicopter touched down softly with around 92% engine and rotor RPM. Once contact with the ground was made, the helicopter became airborne and spun 180 degrees to the right. Unable to maintain a hover, the helicopter softly touched down once more, but due to extremely rocky terrain, it rocked back and forth before it settled. Due to the aggressive slope angle the helicopter was now resting on, I knew that by lowering the collective would result in a static rollover and destruction of the helicopter.

I lowered the collective very slightly and rolled on the throttle to obtain 104% engine and rotor RPM. With the left skid pinned into the slope, I lifted the helicopter to a hover and departed the hillside. Once airborne, I flew the helicopter to the nearest safe landing area where I conducted prescribed pre shut down procedures.

Once the rotor blades had stopped, I secured the master battery switch and fuel valve. Upon inspection of the helicopter, I noticed the right passenger side of the tail boom had a crease of about 4 inches long by 1 inch deep where it mounts to the upper fame. Upon field inspection from our mechanic no other damage to the helicopter has been found at this time.

NOTES:

POWER LOSS AFTER TAKEOFF
CALLBACK

R-44, Commercial, Phase: Cruise, Name Withheld, Jul 2009

The flight, which took place well before sunset, involved three persons (pilot and two passengers) with normal lift off, climb and initial cruise.

Approximately 4-5 minutes into the flight, cyclic and collective controls were noted to be abruptly more difficult to manipulate with particular difficulty lifting the collective to a power setting of greater than 15 inches or manifold pressure. Hydraulic toggle switch on the pilot's cyclic as well as the hydraulic circuit breaker were noted to be in normal/on position. No abnormal indication were noted on the warning light annunciator panel.

A turn was made back towards the airport.

All instrument indications (oil temperature/pressure - cylinder temperature - engine and rotor RPM) were in the green with normal readings. Most notable was the continuing extreme difficulty actuating the collective past the above described power position as if it was mechanically restricted. Below that position, it moved freely.

Flying southerly now above the beach, it became obvious that the

15 inches of manifold pressure power setting with current load conditions would not maintain altitude and airspeed with sufficient reliability to safely complete a return to the airport over populated areas so the decision was made to make a precautionary landing on a vacant portion of the beach just north of the extended center line of the departure airport.

A call was made to the Tower to alert them to our problem, position, status or persons on board and intentions. We then executed a modified running landing on the sand because of low power limitations. The helicopter came to a controlled stop and final check of all engine instruments showed normal function. Shutdown procedure was followed in an expeditious manner.

On initial inspection of the cockpit, no obvious abnormal mechanical issues were noted. The aircraft was secured after both passengers were safely escorted away from the aircraft. Recovery operations then successfully followed all appropriate authorities were expeditiously notified.

Lessons Learned:

Be prepared to lose power at any stage of flight.

NOTES:

DISTRACTION EQUALS LOW FUEL
CALLBACK

R44, Passenger, Phase: Cruise, Name Withheld, June 2010

I was flying a series of 0.5-0.6 hour tours.

Customers were being loaded and unloaded without engine shut-down in order to save time and enable us to complete the tours within their time and budgetary constraints, and tours were taking longer than anticipated with some customers asking for longer time around the city.

On the fourth (and last) tour the low fuel light illuminated approximately eight miles out from the airport. I did not have suffi-cient fuel to complete the tour. I had already contacted the tour inbound so radioed to them that I was setting down due to low fuel and found a flat place and set down to call for fuel.

Unexpected tasks in the office that morning (filling in for another instructor on an instructional flight and clerical duties) filled up much of the two hours of preparation time I had allotted myself to prepare for the flight. Rushing through the preflight phase made me more likely to miss the error I had made in fuel calculation. Help from other instructors (because I was busy) involved fuelling up the helicopter for me, removing me somewhat from concentrating on that flight and what was required for it.

Lessons Learned:

As the last group was being loaded I should have done a more complete assessment of fuel quantity and reserves and shut down to obtain more fuel.

I over-emphasized weight considerations, not filling the tanks completely before the tour group arrival in case the passengers turned out to be heavy (they were not).

Extensive experience in the R22, for which 5 gallons is an adequate fuel reserve, and less cross country/long flight experience in the R44, for which that reserve is inadequate, made me tend toward an insufficient estimate of what I needed.

The wish to keep the tours moving along and save my customers time and money made me rush both moving forward with the first tour (when I could have stopped to fill the tanks on realizing that we were not weight limited), and also made me reluctant to stop in the middle of the series of tours.

NOTES:

INCORRECT HYDRAULIC PUMP
CALLBACK

Bell 206LR, Ambulance, Phase: Parked, Name Withheld, Aug 2010

Well, it all boiled down to I was in a hurry, skipped a step and messed up. I noticed oil on the transmission deck. Upon further investigation I determined that the hydraulic pump was leaking. An L4 Hydraulic pump had been ordered for another aircraft but ended up not needed, so it was returned to my inventory. I entered it into the Parts Inventory List but had not had time to put it on the shelf. I saw the pump on the desk, and decided to use it and not pull the other pump off the shelf. (Keep in mind that at this time I did not know that there was a difference in the L1/L3 and L4 pumps.)

I looked at the Bell IPB (Illustrated Parts Breakdown) and noted the P/N's (Part Numbers) for the three o-rings and gasket. I saw two pumps in the illustration but the drawing showed a physical difference, so nothing flagged a problem in my mind since the two pumps I had were identical on the outside. I installed the pump and we (the Pilot and myself) performed an [engine] run-up. Everything started well, but after about ten minutes of run the Pilot noticed a whining noise coming from the upper deck. He shutdown and after checking things out we opted to try another run-up. The second run-up went just as well and no noise was noticed. We figured that there was an air bubble in the pump that had worked out upon shutdown, so with no Operational Check Flight required, we put the bird (helicopter) back in service. I asked the Pilot if he wanted to do a "turn around the patch" but he didn't think it was necessary. The crew was dispatched on a flight around XA:30 the next morning.

After flying about six miles from Base the Medical Crew noticed a whining noise and expressed uneasiness with continuing the flight. The Pilot turned the aircraft around and flew direct back to the hangar. About half way back to the hangar the Pilot was able to hear the whining noise. I replaced the hydraulic pump with the pump on our Parts shelf. During the change, I noticed the difference in P/N's and went back to the Bell IPB and researched the difference in the pumps and realized the mistake I had made. I stopped everything and contacted our Supervisor to see how I needed to proceeded from this point.

The Supervisor contacted Bell and they determined that no damage was done to the aircraft and all I needed to do was to install the correct pump. I installed the correct pump and the aircraft has been working properly since.

Lessons Learned:

Don't just think you know something, check to be sure.

Always check all the parts in the IPB (Illustrated Parts Breakdown), not just the parts that you can't remember the P/N's

Other than my shame and embarrassment, this situation turned out well, this same kind of mistake with other situations may have catastrophic results.

CALLBACK Comment: Reporter stated the Bell 206-L Transmission, Transmission floor pan and reference to the "upper deck" are all located above the interior cabin ceiling of the Bell Long Ranger. The hydraulic pump is mounted on the side of the transmission.

NOTES:

FATIGUE MONITORING
CHIRP

Unknown, Commercial, Phase: Take-Off, Name withheld, Aug 2017

On [ ] morning my colleague and I planned a flight to an offshore installation. The flight planning was all as usual but my colleague looked tired and I asked him if he was feeling ok. He confirmed he was ok but said he had been working several days overtime on his normal days off. I asked why he was doing so much overtime. His answer was, 'I feel pressured to do so as I think the next selection for redundancies will be based on flexibility. I don't want to report fatigued as sickness absence was used in last year's Matrix to select redundancies.'

We did a normal start up and my colleague who was on the radios asked for taxi clearance with ATC. We were cleared to taxi and hold at a normal helicopter holding point. There was a little morning rush where a few fixed-wing aircraft were ahead of us before it was our turn.

After several minutes of holding (in a sterile cockpit) Ground asked us to switch to the Tower frequency. My colleague did not respond to the call, and when I looked across I noticed he was asleep.

I answered the call, which woke my colleague up. He replied to Tower and followed normal procedures. We continued the flight as normal.

Lessons Learned:

CHIRP Comment: Severe commercial pressure for offshore helicopter operators and the threat of redundancies for their staff have become routine. The veracity of the threat of redundancy in this case is unknown but the reported pilot's perception that demonstrating flexibility through working overtime was material. It should also be noted that overtime attracts financial rewards; newcomers to the industry in particular might not appreciate the possibility of cumulative fatiguing effects.

The problem of pilots pushing themselves beyond sensible limits may be compounded by the unwillingness of colleagues to take action. In an earlier redundancy round (with a different operator) one Captain described how some of the First Officers he flew with appeared to be unfit/stressed/distracted and that he kept an especially close eye on them. He may have spoken to them informally and offered advice but he didn't say that he had reported his concerns to the operator.

The bottom line is that we need to acknowledge that pilots are not pieces of machinery. They are human beings and when faced with prolonged uncertainty and the threat of redundancy, with all that means for themselves and their families, their judgment and decisions may not be based entirely on professional factors.

NOTES:

CHAPTER 2
TORQUE & POWER

"Like all novices we began with the helicopter but soon saw it had no future and dropped it. The helicopter does, with great labor, only what the balloon does without labor, and is no more fitted than the balloon for rapid horizontal flight. If its engine stops, it must fall with deathly violence, for it can neither float like a balloon nor glide like an airplane. The helicopter is much easier to design than an airplane, but it is worthless when done."

Wilbur Wright

POWER ISN'T EVERYTHING
CASA

Bell 206, Commercial, Phase: Hover, Name withheld, Mar 2014

I started flying when I was fourteen. My goal echoed that of any young pilot at any flying school around the world who had been bitten by the aviation bug — work my way up to flying the biggest aircraft I could to the furthest reaches of the Earth. However, after eight years flying aeroplanes my goals were turned on their head after I experienced my first flight in a helicopter. I was awestruck! The freedom of landing and taking off from anywhere, no wings, no runways, hovering... I was hooked. I signed up to a course and after two years I was a fully qualified helicopter pilot working in south-east Queensland.

In 2008 I had finished my training on the Bell 206 Jet Ranger and was keen to get out and use my new qualification on an all-powerful turbine helicopter. An early opportunity came in the form of an air-to-air photography job. The job seemed simple: follow three aircraft around the ranges, filming them for a training video for the government. I had another pilot on board to handle the filming, an ex-helicopter instructor with thousands of hours on the Jet Ranger, who was now off flying light jets for another company.

The day was beautiful. CAVOK was forecast throughout, with light surface winds. The area forecast mentioned that the winds would increase in strength quite sharply with height but it wouldn't be an issue as most of our work would be at low level. We departed from the airfield behind the three aircraft and the task went exactly as planned. After thirty minutes of filming, the last sequence involved the three aircraft landing in a pad together and then departing for lunch at home.

The cameraman told me to get up to 1000 ft AGL and hover in place so we could get a better view of the activity. I manoeuvred the helicopter into the spot and waited for the aircraft to do their thing. After only about a minute of hovering my whole world suddenly fell apart. I felt a strong gust through the open door and the aircraft immediately started to turn to the right. I applied left pedal to keep the nose straight, but it was taking more and more pedal to keep straight. About a second too late, I realised that I had run out of pedal but the aircraft was still turning right.

At that point the aircraft had turned thirty degrees and was accelerating its rotation. After about another second, it was acting like an out-of-control rodeo bull, my windscreen was a half blue half brown blur, and I had no yaw authority left. My only reasonable option was to get some forward speed and fly out of the spin, so I jammed the cyclic to its forward stop and lowered the collective. My windscreen was now full of trees and drought-bare Queensland hills. If this didn't work the only way left to oppose the rotation was to shut down the engine and enter autorotation — something I didn't fancy in the slightest at this early a stage in my career as a helicopter pilot.

After what felt like a lifetime the aircraft gradually started to respond and the forward speed crept up slowly. As it did the aircraft ceased its out-of-control spinning and dished out of the dive at 250 ft AGL.

As I climbed the aircraft back to height I had to fight to regain my situational awareness. I established us clear of the other three aircraft and checked the helicopter's systems, which fortunately looked just fine, then asked my passenger if he was okay. He told me that he was, adding that he had caught a chance glimpse of the torque gauge during the initial onset of the rotation and it had been indicating an over-torque. The density altitude (DA) we had been hovering at obviously required a higher-than-usual power demand. My addition of full left pedal and extra collective had resulted in a minor over-torque.

I informed the other three pilots what had happened and landed in a paddock to have the helicopter inspected prior to returning home. While we waited for the engineers my passenger and I debriefed the incident. We both thought LTE (loss of tail rotor effectiveness) was a probable culprit. We were only at 1000 ft AGL but the density altitude was significantly higher, the wind was gusting and veering at height, and we had full fuel and crew. The power demand on the aircraft in the hover would have been quite high. Even though the aircraft remained undamaged I was upset with myself for putting it and us in such a vulnerable position. The whole scenario could have been avoided by flying a slow-speed orbit, but I had invested too much confidence in my 'all-powerful turbine machine.'

Lessons Learned:

The lesson I took away from that day was very simple — you are taught basic airmanship techniques (nose into wind, know your performance, watch your power) for a reason. Imagining your aircraft has the power to overcome poor decisions is a mug's game.

I was lucky we had height that I could trade for enough speed to fly out of the rotation. After reading many accident reports since then I realise that a number of Jet Ranger pilots have not been as lucky.

NOTES:

FAIL SAFE
CASA

Bell 205, Commercial, Phase: Cruise, Lloyd Knight, Aug 2010

It seems to me that often pilots do not understand the principle of failsafe design, as it applies to electrical/electronic control of aircraft systems. To illustrate this, I will describe an incident that almost had a nasty outcome involving the operation of the hydraulically boosted control system in the Bell 205 helicopter.

Because of the heavy forces needed to control the rotor system, a transmission-driven hydraulic pump supplies pressure to servos that reduce the stick loads felt by the pilot. In the case of total hydraulic failure the helicopter can still be flown, although with some difficulty. Because hovering in this condition would be virtually impossible, a run-on landing would be required.

A more difficult failure may occur when one hydraulic servo fails, but the others continue to work. This means that the controls are boosted in some parts of their movement, but not in others. Such a failure could easily result in an aircraft that is 'unflyable' by the average pilot. Bell therefore provides a switch allowing the pilot to disable the hydraulic system. The pilot still has to contend with a total hydraulic failure, but all the stick forces are equally high, and the aircraft is still flyable. The hydraulic disable system is failsafe. This means that an electrical circuit is used to hold the hydraulic system in the disabled condition. When the hydraulic system switch is in the 'on' position, this circuit is switched off and the hydraulic boost is switched on. Likewise, if the electrical system fails, this circuit will be de-energised, or off, and the control linkages will continue to be boosted, regardless of the position of the hydraulic override switch. This prevents loss of the aircraft electrical system from causing a total hydraulic failure. In short: if electrics 'off', then hydraulics 'on'; for hydraulics 'off', electrics must be 'on'. I was returning from an offshore sortie one day when the pilot of another aircraft called on the radio, in a highly agitated voice, that he was losing control. He said the hydraulics kept cutting in and out, and the aircraft was rolling and pitching violently. There was real panic in his voice and I could hear his passengers shouting in the background. Another pilot called, 'Switch off the hydraulics'. He responded with, 'I've switched off the hydraulics, and pulled the circuit breaker, I think we're going in'. I called out as calmly as I could, 'Leave the switch in the off position and push the circuit breaker back in.'

After a minute's silence he came back with, 'I did that and I have control back with no hydraulics.' What he had done by pulling the circuit breaker was negate the override system by de-energising it, which was the same as turning the hydraulic system back on. Pushing the circuit breaker in turned the hydraulics off again. He proceeded back to base and made a run-on landing on the flight strip beside the runway.

Lessons Learned:

Follow the flight manual procedures. Do not apply your own overkill additional actions.

NOTES:

LIVE, LEARN, SURVIVE & BE HAPPY
CASA

Unknown, Commercial, Phase: Landing, Name withheld, Aug 2012

I hadn't planned on writing yet another 'close call' story – after all, my experiences are probably similar to everyone else's – but there really isn't a better way of illustrating how my attitude to risk in flying has changed over time. So, I've included a few brief stories at the end of this article as examples of lessons learned or mistakes I wish I'd never made.

Back in my bush flying days, it seemed the list of things that could kill me was almost endless – overloaded machines with barely adequate performance, lousy weather, the kind of territory where an engine failure inevitably meant disaster, dodgy maintenance, indifferent company management etc.

I eventually became inured to these everyday risks, and a fatalistic attitude set in. I used to think to myself: Well, if one thing doesn't get me, something else probably will, so what's the point of even trying to manage anything? Besides, I'm fireproof and it'll never happen to me anyway, so why worry? Just press on and hope for the best. This went on for years and somehow I survived, but some of my colleagues didn't. It gradually dawned on me that, if I wanted to live, I'd better start managing all the risks I possibly could. I mean, how long could my luck last? Sure, there were plenty of things I still had no control over, but (when I thought about it) I could influence a surprising number, for better or worse.

So, when I was next faced with situations outside my comfort zone, I either adjusted things until I felt the odds were mostly in my favour, or I declined the task altogether. If pressured by my employer to continue unsafe or unduly risky practices, I quit. I lost a few jobs that way, but it didn't do me any harm in the long run and, perhaps more importantly, I'm still around to talk about it.

Since those days, I've learned that feelings of invulnerability, hopelessness or resignation are recognised hazardous attitudes that can be overcome. I wish I'd known that beforehand, instead of belatedly discovering them for myself, but better late than never, I suppose.

The first story concerns fuel – or lack of it, to be precise. In the interest of satisfying my employer's or my customer's demands for max payload, I used to fly without legal and/or sensible alternate fuel for weather diversions. I figured I would always make it to my destination, either because I knew the area well and felt I could safely bust the proper procedures (like scud run), or because I had some 'homemade' instrument approaches worked out if conditions were really bad. Actually, I always did manage to make it (although sometimes with only fumes in the tank and my heart in my mouth) but, looking back and thinking about the risks I ran in those days – for no good reason – now makes my blood run cold.

Still on the subject of carrying max payloads to please the boss, I've lost count of the number of times I've squeezed out of tiny take-off areas and missed obstacles on climb out by the skin of my teeth. All in a day's work, you might say, but the margin for error really shouldn't be zero...

I recall one occasion when my task was to land a heavy load of passengers on a ridge-top pad. From prior experience, I knew the helicopter's performance would be marginal at best, but I pressed on regardless, not bothering to carry out a detailed assessment of the approach, or to consider other options (such as landing elsewhere and making the passengers walk). The upshot was that I ran out of power on short final, exceeded engine and transmission limits and touched down rather firmly on the pad with the rotor low-rpm horn blaring and the collective up around my armpit. A narrow escape... but why did I do it?

Lessons Learned:

I think the three factors mentioned earlier could be relevant to these (probably not uncommon) incidents:

- invulnerability ('I've done this before')
- hopelessness ('The passengers expect me to land there')
- resignation ('My job is on the line if I don't do this')

These days, I do my best to be consciously on guard against potentially hazardous feelings such as this, as part of my intention to live a long, happy and safe flying life.

NOTES:

HELI-SKIING
CASA

R22, Private, Phase: Landing, Kris McLean, Apr 2009

I started flying in sailplanes thirty years ago, wondrous flying machines which can go great distances without an engine. From there I moved into powered fixed wing: C172s, Warriors and the like.

Eventually I got into helicopters. I soon learned they are flying machines of a different pedigree altogether; there are a lot more moving parts and they are harder to manage. They are also easier to overload: dual in a two-seater with full fuel in summer cannot only make you illegal, it can make it impossible to get airborne. Turning is different too; left pedal turns in American-built machines are accomplished with increased tail rotor pitch and consume more power than right pedal turns. Despite this, left turns are generally preferred as they are less inclined to accelerate than right pedal. Furthermore, rotary wing aircraft can be easier to tip over than planks, so much so that the training syllabus has a section on not doing it. 'Avoiding dynamic rollovers' it's called.

After nine hundred hours in my little R22, most of it at 500 ft, I finally decided I knew enough about the vagaries of the helicopter to try my hand in Australia's bit of high ground, the Snowy Mountains. With a max altitude below 7000 ft, the destination seemed low, so when flying with my sibling presented itself, I grabbed it.

The big day was fine but blowy, so I was careful to land into wind because two up in a 22 at six and half thousand feet don't leave much in the way of that great saviour of average pilots — surplus power margin. We had a ball, checked out all the Victorian ski fields, then filled up at Hotham and did Kosciusko too. We were on the way home when I got the idea for just one more picture. What is it about human nature that makes us greedy? We'd had a great, safe day's flying, but I had to push my luck for one more photo opportunity. As we approached up-slope to a hilltop, the snow was heaped up into a long drift that looked steeper the closer we got to it. The wind had dropped during the day to just five to eight knots, so I decided to turn down wind and try for some more level ground.

Instinctively, I turned left. Big mistake, the low-rotor RPM horn came on mid-turn, and by the time I'd got us pointing down slope, the machine settled with power at 35 knots and started to slide. I tried to wind on some more throttle, but the governor had already maxed it out. Slipping downhill, I racked my brain for some ideas on what to do next. It occurred to me that there hadn't been a lesson in my training on what to do when careening down a steep, snow-covered slope in a helicopter. Clearly I needed to do something before I ended up with a bent whirly-bird. I attempted to raise the collective. The low rotor RPM horn came on again, so I tried lowering it. The skids started to dig in and it felt like we might nose over. Finally, in desperation, I kicked in a boot full of left pedal. The bird slewed, slowed immediately, but started a dynamic rollover to the right. I actually breathed a sigh of relief. For the first time in the last five, terrifying seconds, I knew exactly what to do. I floored the lever and pushed in as much up-slope stick as I could.

We plopped back onto two skids and and I looked anxiously over to my passenger.

'New slant on heli-skiing, hey?' he said and flashed me a wan smile. I felt like a wrung-out dish cloth, but I knew we couldn't stay there; we had a nasty down-slope list and it would be dark in a few more hours. I booted junior out and handed him the mobile phone in case I came to grief extracting myself from the predicament I'd put us in.

I took a few deep breaths, calmed myself, then raised the lever and wiggled the pedals to free up the skids. The bird unstuck clumsily and I had cyclic everywhere till I got a hover. Next I tried for some out of-ground-effect manoeuvring to check there was nothing wrong with the engine. Lastly, I put down facing into wind on the edge of a precipice so I could nose over after I picked up my passenger. The rest of the day was uneventful.

Lessons Learned:

So what's to learn? I needn't have taken on full fuel at Hotham, and I definitely shouldn't have used left pedal, or turned down wind at altitude.

I should have shut down, got out and given the helicopter a thorough inspection before taking off again. I'm very grateful to have escaped the episode unscathed.

All I can say is, if you take a chopper to the snow be careful, do a reserve power check and always land into wind, because it's no fun scaring the daylights out of yourself and your passengers.

NOTES:

FREIGHTFUL FLIGHT
CASA

Bell 212, Medevac, Phase: Take Off, Name Withheld, Oct 2009

Twenty years ago, I was the base manager of a helicopter operation supporting an oil drilling rig in Bass Strait. We flew a couple of Bell 212 helicopters, and were based at Welshpool and the semisubmersible rig the Zapata Arctic, which was about 100nm to the south east. For some strange reason not important to this story, we flew with two pilots aboard Monday to Friday, but at the weekends we were permitted to fly single pilot.

One lovely cloudless Saturday morning, I was taking it easy when the telephone rang. 'Hello Ron,' said the oil company representative, 'We have a medevac. There is a guy on board with a fever and the medic thinks he should be evacuated. Can you do it?' 'No worries!' It was a beautiful day to go flying. No cloud and a light and variable wind.

One and a half hours later, I landed on the huge deck of the Zapata Arctic. It was standard procedure not to shut down, but to keep everything burning and turning, with the throttles at ground idle. Soon enough there was activity on the helideck, and the helicopter landing officer (HLO) escorted three men on board. One of them was in obvious distress and the other two were there to look after him. It was not uncommon when a medevac took place for all the patient's mates to volunteer to look after the guy. A free night in town was on the cards. I watched them being loaded in and secured. I was vaguely conscious of activity to the rear of the helicopter and assumed the deck crew were loading luggage in the tail boom locker.

Eventually the passengers were securely strapped in and the sliding door was slammed shut. I looked round at the passengers and gave them the thumbs-up. One of them returned the gesture. There was no voice communication available between the cockpit and passenger cabin. The HLO stood outside the cockpit window and handed me the manifest. It simply had the names of the three passengers on it, plus the fateful words: 'some freight'. In those days standard weights were used for passengers, and everyone was a bit lax about weighing freight if the helicopter was light. In this instance I only had the three passengers and about 1,000 lbs of fuel and I assumed the freight was luggage. 'Piece of cake!' I thought to myself.

When the deck was cleared I lifted into a hover to check temperatures and pressures (Ts and Ps) and power availability. Everything was in the green and I was only using about 75 per cent torque. I moved to the edge of the deck for the takeoff, pointing west for home. A final check and everything seemed normal, except the aircraft seemed a bit tail heavy. Ah well, it was probably that the direction of takeoff I had selected was a bit down wind. At this weight, no worries! I pulled close to full power and up we went like a homesick angel. But as soon as I left the ground cushion, the tail really dropped. I shoved the stick forward, and to my absolute surprise and fright, I could only move the stick about three cms forward before I hit the stops. By now the helicopter was climbing fast and clear of the deck, but the tail was still dropping, and the nose was an alarming 30 degrees up and rising. Terror took control. The nose continued to rise and the airspeed was minimal. I guessed I was clear of the oil rig and I did what seemed to be the only action available. Instinctively I rolled steeply to the left, away from all obstructions and applied a hefty piece of left pedal. The aircraft banked hard left. It seemed about 90 degrees of bank, but was probably only about 60, and the nose swung down through the horizon. Immediately the airspeed started to rise, but the sea was getting really close.

I rolled level again and pulled out of the dive, but this time, when stabilised, the nose was only about 10 degrees up and the airspeed was building. Soon I had 60 knots on the clock and the rear elevator started to assist the attitude control. I still had the cyclic jammed hard against the forward stop, but at least the nose was in a constant attitude and we were gently climbing. I had a stabilised airspeed of about 80 knots and I had reduced power to about 68 per cent torque.

We had survived! Now I had to try to sort out the problem.

There was obviously something seriously wrong with the configuration of the helicopter. I triggered the radio, attempting a controlled voice, but it was more likely close to a scream, and asked the oil rig if they would mind educating me as to what the 'freight' was. Their reply floored me. 'Captain, we loaded 200 kilograms of core samples into the tail locker.' That explained a lot. The tail locker was only cleared for a maximum of 400 lbs and then only when the front of the helicopter was fully loaded. The centre of gravity was so far behind it was probably somewhere near New Zealand. It was foolhardy to even think of getting back on the deck of the rig. I reckoned that as soon as the airspeed dropped below about 45 knots, the elevator would be useless, and the nose would start its horrible 'pointing-at-God' routine. I looked behind at the passengers. Three extremely white faces with huge, round eyes stared back. But these were oil rig workers and not easily scared. I pointed at the biggest bloke and indicated that I wanted him to climb over the back of the front row of seats and get himself into the vacant co-pilot's seat. He nodded and gingerly unstrapped. Slowly and carefully, he removed the headrests and hoisted himself over the obstruction and into the co-pilot's seat. There was an immediate easing of the centre of gravity problem. It's amazing what 300 pounds of oil rig worker can do! The nose dropped a bit and I was able to ease off full-forward cyclic. I then indicated that the other two guys were to move into the most forward seats. This they did, and there was a further easing of the problem. Now I could achieve about 90 knots with the nose more or less level with the horizon and have a bit of forward cyclic to play with. We gingerly made our way back to Welshpool with me quietly giving a special thanks to the designer who thought to include a moveable elevator in the Bell 212.

On arrival at base, I decided not to let the speed drop below 60 until I was just off touchdown. Welshpool had a lot of grass areas edging the runway and I decided to run it on over the grass. This was accomplished without too much trouble, and soon we were sitting on the ground surrounded by anxious and curious spectators, all wanting to know what the trouble was. With my most casual voice I asked the chief engineer to remove the offending freight from the tail locker. When that was done, I was able to lift off and hover normally back to dispersal.

Lessons Learned:

Since those days, things have improved. Standard weights for passengers are no longer used. All freight is properly weighed and notified to the crew – even when the aircraft is largely empty.

However, one thing remains. All helicopter pilots should be properly trained in aerobatic manoeuvres.

<u>NOTES:</u>

CHAPTER 3

WIRES

"The whole scene was a helicopter's nightmare, because everything was basically wires or tower."
Dave Cooper

TOO CLOSE FOR COMFORT
CASA

Bell 206 LR, Fire Fighting, Phase: Landing, Ross Knudsen, Oct 2012

All too often we read or hear accounts of helicopters experiencing near-misses or collisions with power lines. If the crew survives to tell their tale, their explanations of these events are many and varied. We all know there are numerous factors including weather, mechanical problems, pilot/crew error, fatigue , etc. However, I always thought my training and vigilance in this high-risk environment would never let a near-miss or collision occur on any of my flights.

I had been deployed with my pilot to assist with firebombing duties as an air attack supervisor (AAS) on an active fire in the Gypsy Creek area of the Bunyip State Forest east of Melbourne. I was an accredited AAS with ten years experience in both rotary and fixed-wing aircraft. Our working platform at this time was a Bell 206 Long Ranger. Training included briefs on hazards and power lines and safety was always a priority-for good reason. The helicopter was mechanically sound, the pilot and l were it, healthy and hydrated and the weather conditions on the day were hot and sunny, with a moderate wind and good visibility.

The early autumn weather continued to be dry and the regular weather changes resulted in little if no rain. The fresh northerly wind that drove the fire for most of the day abated to calm conditions by early evening. The smoke from the fire settled into the valleys of the ranges and fire behaviour became quite sedate. Firebombing operations ceased by last light and we were instructed to land at Noojee and rest there for the night before continuing operations the following day. Not only had the day included firebombing, but also the plotting of the fire perimeter and reconnaissance required by the Incident Control Centre (ICC).

The following morning our first task involved intelligence gathering about the fire's behaviour and condition, mapping the new fire perimeter and reporting that information back to the ICC. Overnight, the fire had spotted over a bulldozed firebreak along a ridge and was burning slowly downslope into steep inaccessible terrain on the southern flank of the fire. We concentrated our efforts in this area as it was the only active fire perimeter. We used Helitack (a helicopter-delivered fire resources for initial attack on a wildlife) to assist in suppressing the active fire edge. This technique is often very concentrated and intense.

Private property bordered the state forest directly below this ridge and consisted of open, undulating terrain, with small vegetated areas. Cattle grazed on the grassland and a farmhouse was located up on a ridge close to the fire perimeter. During our operations, we had flown over and close to this house on numerous occasions.

Running east-west and downhill of the house was a single-strand power line. Being silver in colour, it was quite easy to see. The supporting timber poles were also clearly visible, as they stood alone on the open ridges. Another span ran from one pole up the ridge to the house. The pilot and I recognised the existence of the poles and power lines and maintained a safe distance at all times.

Late in the morning on the second day of operations, I had a call of nature. I asked the pilot to find a suitable spot to land so I could get out and relieve myself. An obvious level location to land the helicopter was on the creek flats a few hundred metres downhill from the house. Visibility was good and there was no turbulence in the lee of the range. We descended following the ridge, passed over the silver power line to the flats and came to a hover about 10 metres above the ground. The pilot then taxied to the left and towards rising terrain between two ridges at 10 knots ground speed. The silver power line was clearly visible up and away from us.

Then - power line!

The pilot and I saw the power line at the same time and a shiver pulsed through my body. Where did that come from? The power line was now under the rotor disk and just above the cabin. Pull-up! I could not believe how close the rotors came to the power line and possible wire strike. Only the skill of the pilot averted disaster by pulling up and manoeuvring away from danger 'That was much too close'. Apart from the pilot's skill, the only other thing that saved us was the slow forward speed of the helicopter.

The pilot quickly found a suitable spot to land and I jumped out. We looked at each other realising just how close to calamity we had come.

The power line we almost collided with was not the one we had identified earlier. This was a separate span, black-insulated, quite narrow and running parallel to the silver strand, but further down the hill. It was almost invisible and had sadly slipped through our 'vigilance and situational awareness net'.

Once airborne, we followed the black power line to see where it went. One thing that made it difficult to identify was that its supporting poles were located in stands of trees growing on the ridges, with the long span drooping low into the valley it traversed. We hadn't anticipated or expected another power line running in close proximity and parallel to the other one. It was a potential trap for anyone!

Lessons Learned:

This was a really close call and a disturbing incident that could have resulted in severe consequences. It highlights the importance of vigilance and the need for constant visual alertness when operating at low levels in unfamiliar terrain, particularly in helicopters. These are basic principles of operating safely.

<u>NOTES:</u>

WIRE WORRY

CASA

Bell 206 LR, Commercial, Phase: Landing, Richard Guay, Dec 1995

December 1995 – a beautiful day at Jackson's International Airport, Port Moresby, Papua New Guinea (PNG) with almost perfect flying weather forecast.

I was the base manager/pilot for one of the larger helicopter operators in Papua New Guinea and that morning we had a government charter, including a local government minister, a few officials and a police escort. This was a full complement of six passengers for my Bell 206 LongRanger. The charter was to Waitope Village, and scheduled takeoff time was 0700, something I failed to notice during the sequence of events for that morning.

All passengers arrived on time, so after weighing them, and manifested, we actually departed on time. The route took us to the north of Port Moresby into the beautiful Waitope Valley. Waitope Village is located at the top of the valley, at the base of Mt Albert-Edward, in the Owen Stanley Mountain Range, at about 5000 ft altitude. The airstrip is suitable for small fixed-wing aircraft, with a tourist lodge near the end of the strip.

As I lined up my approach to the far end of the Waitope airstrip, still about half a mile out and about 500 ft AGL, we passed a primary school on our left, which the local government minister, the member for Goilala, told me he attended as a child. This prompted him to ask me to land at the school, rather than the airstrip, as he was due at the school to address a grade six graduation. Of course I complied, and immediately commenced a descending left turn from the eastern boundary of the school to initiate a high overhead inspection of obstacles and terrain to select an intended landing point.

There was a large soccer field in the middle of three school buildings, with a basketball court on the eastern side of the complex. I chose the soccer field.

As I neared the southern boundary of the school grounds approaching from the east, I commenced a right-hand turn surveying the layout of the school and possible forced landing areas for the approach. I noticed three large high-tension wires on large power poles approaching the school from the south. My eyes followed the wires, which turned east and literally formed a border along the soccer field's northern side. I also noticed three steel poles: two forming a ninety degree turn from south to east on the southern edge of the school, with another pole approximately 130 metres away on the northern side of the soccer field. The wires appeared to terminate on the second pole on the southeastern side of the soccer field, while the third telephone pole on the northern side of the soccer field appeared to have no wires attached.

Rather than levelling off at about 200ft AGL and conducting a full 360-degree inspection of the site, I opted simply to continue the approach resulting in a 270-degree flight path, executing my approach to the east.

I flew between two buildings, commencing the approach by slowing my airspeed and continuing a standard approach and resultant rate of descent to the field. I kept a close watch on the heavy high-tension wires immediately to my left, clearing them by about six metres from my rotor tips. Just as I shifted my field of vision ahead to my intended landing point, I saw a smaller wire literally at the end of my HF antenna (only about two metres in front of the nose of the helicopter), perpendicular to my aircraft, and level with the antenna itself. This indicated the wire was actually under my rotor disk, and I still had translational lift (about 20 knots of airspeed) and a four to five degree approach angle.

At this point everything I did was by reflex. I hauled back on the cyclic, pointing the nose of my B-206 straight up, while holding my approach power with the collective. This manoeuvre allowed me to drop straight down, tail first, before reaching the wire and then level the chopper in one continuous motion. I then pulled enough collective to try to arrest my descent, and almost succeeded. However, the rate of descent was too rapid for such a low altitude and I hit the flat surface of the playing field rather smartly. The skids spread out completely flat and there we sat. I was surprised – I couldn't believe I had actually 'pancaked' the skids.

The emergency locator transmitter (ELT) was howling so loudly in my headset I could hardly hear myself think. The rotor was still powered and turning at full flight RPM, and the fuselage was now about a metre closer to the ground, with the rotor spinning at about neck height for the average person. I asked my front seat passenger if he was all right, and he replied he had not felt the impact, and neither had I. The surprised look on his face and his bulging eyes told of his real state of mind. I then realised that at any moment one of my passengers in the rear compartment could bolt from the aircraft, probably stand up and start running with most likely terminal results.

I snapped off the ELT, reached down and flipped the fuel emergency cutoff switch, silencing both the beacon and engine in a matter of seconds, and said to my passengers (who still had their head sets on), 'Listen to me! Do not leave the aircraft! The rotor is low enough to knock your head off. Exit the aircraft slowly, bend way over, and walk away. Make sure you do not stand up until you are well clear of the rotor.'

Fortunately all of the passengers did as they were instructed.

Thinking back, I am very glad I gave the 'boring' pre-flight briefing to the passengers, telling them to follow my instructions 'in the unlikely event of an emergency'. I can only guess that was why they didn't dart from the helicopter immediately.

Luckily, no one was injured, as the rate of descent had been slowed considerably, as well as the shock-absorbing effect of the skids collapsing provided adequate protection. The rest, as they say, is history.

Lessons Learned:

I never made another approach to a village or field again without doing at least one 360-degree over fly, continually searching for wires, regardless of where I was. Sometimes I even did two or three high and low overhead flights if I had any doubt. The mindset of 'saving the passenger money' is not the right one for a safe approach.

Contributing to this accident was the fact that at altitude, I was getting the full easterly glare from the sun in the direction of my approach, but the sun had not risen high enough over the mountain range to illuminate the ground or the wires below me. There were three wires crossing the soccer field from the steel pole at the southeast of the field, across the field to the northernmost steel pole, where they actually did terminate.

The wires had been placed there years ago, and were the result of an abandoned hydro-electric scheme. Believe me, there are very few wires in rural PNG. When I exited the aircraft and looked up, the wires were over the mast of the helicopter. That was a close one.

NOTES:

TELEPHONE LINE FLEXES MUSCLE
CALLBACK

Unknown, Commercial, Phase: Take off, Name Withheld, Jan 2012

Professional helicopter pilots spend many hours honing the skills needed to perform confined area landings and takeoffs. These skills are crucial to the completion of many helicopter missions. Pilots in a multi-crew helicopter operations often use ground observers to help judge clearance from wires and other hazards to spinning rotor blades. But at times even this precaution is not enough to prevent an incident.

The mission was to take up a photographer to take aerial pictures of the scene of a shooting. A landing was to be made near the incident to pick up the photographer. The area was an urban environment with numerous wires. After an aerial reconnaissance was completed, a landing was made to a suitable spot.

After landing, the pilot got out of the helicopter to better inspect the wires he would need to negotiate on departure. There was a single strand wire crossing north to south between two telephone poles. The departure was to the west. Below the wire was a chain link fence approximately 12 feet high. With the way the helicopter was configured, the pilot was not sure he would be able to clear the wire on departure. He decided the best option would be to fly slowly over the fence and under the wire. Beyond the wire the departure area was clear. The estimated distance between the top of the fence and the wire was 20+ feet. The wind was out of the west at 10 knots.

The observer was left on the ground so the load would be lighter. The pilot brought the helicopter to a high hover and slowly moved over the fence. The pilot looked up to check the clearance of the wire again and it looked good, so he proceeded forward. A loud snapping sound was heard and the aircraft began to shudder. The pilot continued forward and brought the helicopter to a landing beyond the fence.

The observer that was left on the ground said that as the helicopter moved over the fence it appeared to have plenty of clearance. Then as the rotor system was just about centered under the wire, the wire began to oscillate up and down eventually coming down far enough to strike the top of the rotor system... The pilot simply did not imagine at the time that the wire could be affected to that extent by the rotor wash.

Lessons Learned:

After the incident, the pilot talked to other pilots in the unit about the incident. Most said they would have done the same thing and had never considered that something of this nature could happen. The exception was a military trained pilot who said that when the military trains for under-wire flights, the minimum clearance is 200 feet.

NOTES:

WIRE STRIKE
CASA

Squirrel, Fire spotting, Phase: Landing, Ward Bremmers, Mar 1998

Helicopter pilots are all aware of the danger that powerlines represent. The nature of the work means that somewhere, sometime, each one of us is going to have a close call with a wire.

Throughout my flying career I have considered myself very mindful of powerlines and wires in general. I have known several pilots who have made contact with wires. Some of them are tragically no longer with us. In the past, as an instructor, I have tried to impress on my students the danger of powerlines and the high level of vigilance required.

In my current role with *Helicopter Resources* I try to impress upon clients travelling with me to watch for wires, and to call any that they see which could pose a possible threat.

"Don't assume that I have seen them," I've heard myself say many times. When a passenger calls out a wire, I make a point of thanking them regardless of whether I have already seen it or not. When landing in an unfamiliar area I reinforce this practice by asking my passengers to once again watch for and call any wires they might see.

Saturday 17 January 1998: The temperature was around 38 degrees with a blustery northerly wind of around 30-35 knots. A wall of flame was converging on a dozen or so houses in a densely wooded area to the south-west of Hobart, Australia.

With me on board the Squirrel were two officers of the Tasmanian Fire Service directing operations from the air. We had already made numerous landings on various properties to advise people of the impending fire front and appropriate action that needed to be taken. On each landing powerlines were a major concern and the crew were told again and again to look out for and call wires, which they did.

At approximately 19:30 in the afternoon the officers on board became concerned about a building facing immediate threat by the approaching fire front. A water tanker was needed, however communications with the required tanker were uncertain. It was decided that we would land on the highway adjacent to the tanker to inform the crew of their task. The highway had been closed to traffic. The intended landing site was close to the fire. Visibility was reduced due to smoke. A slow approach speed was chosen. All on board were watching for wires.

The first hint of a problem came when the helicopter's nose dipped forward and a positive deceleration was felt. A blue flash was seen along with an accompanying bang. Through the chin window I could see a wire. I couldn't believe it – I'd hit a wire!

I remember thinking, "Should I put the collective down and just get it on the ground? No. I'm still flying and I've still got control."

The wire had somehow caught on and behind the mirror. I had to gently take the aircraft's weight off the wire and then move forward slightly to unhook the mirror and then back off the wire.

As soon as the helicopter was clear, we landed and shut down to inspect for damage. The LED that indicates ELT transmission had activated. I tried unsuccessfully to reset it. All those on board were happy to be in one piece. It was only at this point that we became aware that we had actually broken one wire before getting caught in the second. Within minutes the fire officers were back in the air in another company aircraft. I inspected for damage and elected to fly back to Cambridge.

The damage was as follows:

- Scratch marks on the windscreen from the first wire. These were easily polished out.
- A cracked skylight window where the first wire had contacted the outside air temperature probe.
- A line of paint missing just below the landing lights where the second wire impacted.
- Minor damage to the leading edge of one blade where the broken wire may have touched as it snapped back.

One interesting thing to come out of this was the placement of the ELT antenna. On our aircraft it is located in front of the windscreen near and slightly forward of the pitot tube. When the first wire rode up the windscreen it came into contact with the antenna and sent 240 volts into the ELT. You could only describe what remained as a charred mess. I am sure that my lucky escape was due to the fact that I was conducting a cautious and slow approach.

Lessons Learned:

Ward had a lucky escape. I believe that if it had not been for his cool reaction, smooth aircraft handling and sound airmanship in extricating himself from an extremely difficult situation, then he probably would have joined those less fortunate people who have been caught in wire strike traps.

Some basic tips are clear:

- Don't just look for wires but prove to yourself that they are not there.If this means doing another orbit to confirm

wires do not exist through your chosen landing site, it's time well spent.
- Fly at heights which will let you avoid wires.
- Thoroughly brief your passengers as to their responsibility with regards to sightings of potential wire hazards.
- Slow down in reduced visibility.

Lloyd Knight
Flying Operations Inspector
Moorabbin District Office

NOTES:

WE FOUND THE PHONE WIRE
CALLBACK

Kiowa, Military, Phase: Taking Off, August 2009

Before landing, we did 2 orbits over the field to determine suitability for landing. The landing zone was large, about 175 yards long and 80 yards wide. Based on more than 13 years of military, commercial, and EMS flying experience, this landing zone was very typical for off-airport helicopter operations. There were wires on the north, south and east sides of the field. There were trees on the west side. The topography of the zone was quite level and considering the large size of the zone; I determined it to be a very suitable landing area.

The winds were reported to be south south-east at 8 kts. I planned my approach to the south into the wind. We flew a steep approach. My Tactical Flight Officer (TFO) called me clear over the wires on the left side of the aircraft and I ensured we were clear of obstacles on the right. We landed uneventfully on the south end of the landing zone. Prior to first departure out of the LZ, I turned the helicopter around and hover-taxied back to the north end of the zone to facilitate a departure into the wind (to the south).

When done with the fly over, we landed uneventfully back in the landing zone the same way as described above on the first landing. I again turned the aircraft and hover-taxied to the north end of the field to departure in a Southerly direction. After reaching the north end, I was turning the aircraft around with a left pedal turn. I saw a large portable advertising sign in the field and told my TFO, "I have the sign in sight. I'm bringing the tail to the right. We're clear of the sign."

As I completed the pedal turn, I felt a "thud" briefly. There were no unusual cockpit indications, vibrations, or other unusual handling characteristics.

My first thought was that a stick or branch had been blown upward position underneath or maybe even into the tail rotor. My TFO then noticed a wire dangling from a pole by the road near the sign mentioned above.

I landed and shut down the helicopter. On examination, it was apparent that a telephone wire had made contact with the rotor system.

This small telephone wire was strung diagonally across the north-east corner of the landing zone. Neither of us saw that wire running diagonally across the corner of the zone.

This wire was no factor to the helicopter's landing flight path. Due to the viewing angle of this wire while taxiing northbound with many more prominent wires and trees in the background, spotting this wire was extremely difficult. I believe a contributing factor may have been that I had just moments before, taxied the aircraft in the same manner and did not hit anything. Also, the sign was much more prominent and may have drawn my attention more to it than to the wire above. Also, this wire was smaller than many of the others and the only wire strung diagonally across the zone rather than around the perimeter. Lastly, the background made this wire nearly impossible to see.

It is hard to pick out even after the event and knowing it is there. We were lucky in that there was no substantial damage and no one was hurt or injured.

Lessons Learned:

It may heighten the awareness of others to read this — expect the unexpected.

We are trained to look not only for the wires but for the poles. These two poles, with a wire running diagonally, were an unusual place to have a wire but I for one will look closer and longer for wires in unusual, unexpected locations or where they may be masked from view by either the background or the sun.

NOTES:

CHAPTER 4
PROCEDURE

"The thing is, helicopters are different from airplanes. An airplane by its nature wants to fly, and if it is not interfered with too strongly by unusual events or a deliberately incompetent pilot, it will fly. A helicopter does not want to fly. It is maintained in the air by a variety of forces and controls working in opposition to each other, and if there is any disturbance in this delicate balance, the helicopter stops flying, immediately and disastrously. There is no such thing as a gliding helicopter."

Harry Reasoner

THE VORTEX RING STATE
VRASF

Type unknown, Emergency, Phase: Landing, Name withheld, Date unknown

This story was passed on to us by Claude Vuichard, founder of the Vuichard Recovery Aviation Safety Foundation. The Vuichard Recovery Technique is a technique taught how to escape the Vortex Ring State.

I take my hat off to Mr. Vuichard. I owe my life to his recovery technique.

True story. I am a retired US Military Pilot (US Army helicopter pilot with over 37 years of flying experience, accident and incident free all of these years, I'm retired now).

When I retired (from the military) I started flying EMS helicopters. I responded to a scene with a flight at 2:00 am in the morning, zero moon illumination, and overcast skies. I had my ANVIS 9 NVG's thank goodness. My landing zone was extremely tight and I made three attempts to land with no luck so my last option was to land with a tailwind (not the best decision), I should have landed at another location and had an ambulance take my medical crew members to the scene, but I didn't.

On final approach at approximately 400' to 500' I immediately started to settle in my downwash (Vortex Ring State - VRS). I knew right away what was happening, it was not my first time in VRS. The trees were 150' (tall pines), if I had attempted to recover the old way, by lowering collective (altitude permitting) and flying out of the VRS I would have smacked the trees (98% sure that would have been the end result).

So I performed the "Vuichard Recovery Technique" — a technique I learned six months earlier and BAM! I "immediately" stopped my sink rate in less than 50' and prevented us from balling up the helicopter and everyone being killed.

Thank you, Mr Vuichard for developing this awesome technique and I want you to know that my family and I am grateful that I am still alive.

Vuichard Recovery Technique

1. Increase collective to maximum available power
2. Simultaneously apply power pedal to maintain heading and opposite cyclic (15-20° bank) to get a lateral movement (cross controls) for CCW rotor-systems escape to the right and for CW rotor-systems escape to the left
3. As soon the rotor reaches the upwind part of the vortex, the recovery is completed. Average height loss during this recovery procedure is 20-50 ft depending on the duration of the recovery manoeuvre

To avoid all Vortex Ring State accidents the following is recommended:

- Train ab initio only the Vuichard Recovery Technique and this in the first hours of flight training (emotion drives automatic motion later).
- Equip all helicopter with Instantaneous (e.g. based on GNNS data).
- Equip all helicopter with a string in order to have a correct flow indication.
- Train vortex entry specially without vibrations.
- Train downwind approaches with special emphasis of a maximum rate of descend < 300 ft/min @ low speed.
- Review the procedures in every skill or prof. check.
- All civil aviation authorities should review the training manuals of FTO's (flight training organizations).
- All IFR rated pilots should train the Vuichard Recovery Technique in simulated IMC conditions or in FSTD.
- All helicopter manufacturers should publish safety notices.
- All simulator manufacturers should review the behavior of the simulator in vortex ring state.
- Never use the "auto hover" modes out of flat areas, if the system takes the radar altimeter as reference.
- Be alert with multi-engines helicopters in HOGE (hover outside of ground effect).
- Prohibit backwards CAT A procedures in POH's (pilots operating handbook).
- Amend asap the FAR and CS 27/29. Manufacturers must provide data showing the rate of descent to enter into VRS at various weights and density altitude.

Visit www.vrasf.org to view a demonstration of the technique. The website also contains further information about the Vuichard Recovery Technique.

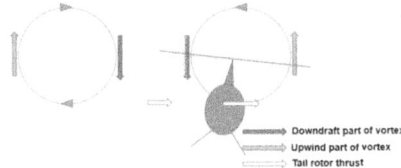

Special thanks to:

    Mr Bruno Bagnoud, Air-Glaciers, Owner and CEO Switzerland, www.air-glaciers.ch

    Mr Hans-Ruedi, Amrhein Valair AG, Owner and CEO Switzerland, www.valair.ch

    Mr Martin Bäbler, AERIALSTAGE®, Aerial Camera Operator, www.aerialstage.com

    Film Realisation: bsv-production GmbH, www.bsv-production.ch

NOTES:

LOSS OF CONTROL AT NIGHT
CALLBACK

Augusta 119, Ambulance, Phase: Takeoff, Name Withheld, Jan 2018

The medical crew and I accepted a call for a hospital transport. I was waiting for an approved flight risk assessment while installing my Night Vision Goggles (NVGs) standing next to the pilot seat.

After the flight risk was approved, I showed it to one of my medical crew and announced green. I placed the iPad on the floor leaning against the center console while securing my logbook, clipboard, and handheld radio.

The medical crew and I ensured the aircraft was unplugged and ready to be moved. The aircraft was pulled out of the hangar, with both of the medical crew as wing walkers, and was placed in its designated position faced into the wind. I completed my walk-around and announced that the aircraft was ready for flight. I climbed in the aircraft and went through my checklist to start the aircraft without issue. The aircraft was safe and ready for flight when I announced to [Company] Communication Center my intentions. I picked the aircraft up after receiving the all clear from the crew. We then cleared the cart, the tail, and the sky before initiating a takeoff.

During my power pull and at an altitude of approximately 40 feet, I felt the cyclic bind while putting in forward left pressure. This caused the aircraft to start a right drift. The crew asked me if I saw the tied down fixed aircraft wing that we were drifting towards and I announced that I had a control malfunction causing me to drift right.

We continued to gain altitude and bank to the right as I was rapidly trying to identify the issue and keep the crew calm by announcing my mental process. I announced to [Company] Communication Center that we had a potential hydraulics malfunction. The bank continued and we started to lose altitude because of it. As we were approaching the taxiway, I was able to free the controls and slow the aircraft down.

I performed a precautionary run-on landing although I did not feel any continued indication of a problem. While doing so the medical crew announced to Communication that we were aborting the flight and that they smelled smoke. I attributed the smell to the run-on landing.

While stopped I started to trouble shoot the malfunction by isolating each hydraulic servo. While feeling out the controls I reached down and saw that the iPad had wedged itself between the center console and the cyclic impeding my ability to move the cyclic left.

I announced to Communication center what happened and briefed the crew to my tremendous mistake.

Lessons Learned:

I am horribly embarrassed by this event as my haste to get the aircraft out of the hangar could have led to a catastrophic situation.

In hindsight, I can see the exact moment where I would normally ensure the iPad is free and away from the controls but had moved onto the next item on the agenda. Because it was night, I did not notice the iPad when I entered the aircraft and again missed another opportunity to correct its placement.

In the future, I will always place the iPad in the same secured location and perform a check prior to start-up ensuring that the controls are visually unobstructed from potential hazards.

NOTES:

NEAR MID-AIR COLLISION
CALLBACK

R44 & EC135, Private & Ambulance, Phase: Final Approach, Name Withheld, Sept 2017

South bound EC135 at 80 knots descending through 700 feet AGL was setting up for landing at a hospital when it encountered an opposite direction Robinson R44 in a head on closing situation. The Pilot in Command (PIC) had just announced "one minute out, secure the cabin for arrival," when he saw a R44 roughly 100 feet below and just to the left at the near 12 o'clock position. Lateral separation was estimated at 100 to 200 yards.

The PIC announced helicopter traffic at 12 o'clock and verbalized turning right and climbing. Both aircraft banked right. The EC135 climbed as it turned right in what both medical crew described as a "firm manoeuvre."

Medical Crew Member seated in the left aft facing seat stated, "That was close!" He then proceeded to provide advisories on the R44's position, altitude and direction of travel.

The R44 cleared low to the left and was seen departing the area to the east. PIC had made up to three advisory calls arriving into the area on CTAF as the EC135 flew from the north to the south descending. No other aircraft announced they were in the area. EC135 had pulse light, strobe light, anti-collision light and position lights illuminated. The EC135's ADS-B and TCAS were fully operational and showed no traffic nor gave any traffic alerts.

As a matter of practice the PIC sets up the GARMIN 650/750 with the 650 on the traffic screen and the 750 on the map screen. Both of these screens would have shown a contact had the R44 been using a transponder.

Following the course deviation and climb to avoid the R44, the EC135 PIC made up to two additional advisory calls on the CTAF. No radio traffic was heard from the R44. The EC135's radios were tuned to the company VHF frequency and [local airport] CTAF.

The aircraft made an uneventful landing at the hospital. ATCC was contacted as well as required company report filed. PIC had the aircraft illuminated with every available light, ADS-B and TCAS were operational and traffic calls on the local airport frequency were being made.

Lessons Learned:

With all the technology available being used it came down to an active scan. Keeping your head on a swivel, ultimately allows for a successful "see and avoid".

Unfortunately, not all General Aviation monitor the local CTAF nor use a transponder as they are not required to in Class G. Hazardous Attitude: "If it's not required. I'm not doing it." Glad this attitude is the exception in the aviation community with the majority talking, squawking and illuminating.

NOTES:

A DECISION BASED UPON SAFETY

CALLBACK

AS 350, Ambulance, Phase: Parked, May 2012

We met in the hangar and prepared to launch. Mission parameters, including possible destination hospitals, were discussed. [X] County protocol dictates that trauma patients with burns are flown to the nearest trauma center. Patients with burns only, are flown to Hospital [X].

At the time the initial call came in, the patient's actual condition was yet to be determined. The first responders had not yet arrived.

When I checked the weather, the radar showed a huge thunderstorm [near] the potential LZ. This storm was part of a large system, moving from the southwest. At the time the call came in, [the area] was covered in a large level 3 to 4 thunderstorm. Hospital [Y] was open. It was also the alternate hospital if [Hospital X] was weathered out. Medics indicated that they wanted to adhere to the protocol and preferred [Hospital X].

At this point, and up until launch, the PIC had neither rejected nor accepted the flight. I had only outlined options and solicited medic preferences. None of the medical crew indicated that I should "just launch and Go to [Hospital Y]." More information was required. It is important to note that the PIC is not directly involved in communications with the [X] County Dispatcher. This is handled exclusively by the medics. The PIC can simply advise the medics, on Go, No-GO, or GO with conditional options. The PIC does not accept a flight. He can only reject a flight if, in his judgement, [Operating Manual] & FAR parameters are not met.

Based on the facts available, I decided that we would need 30 gallons of additional fuel to proceed. Contingency fuel would be prudent in case we encountered airspace clearance problems, had to circumnavigate weather, or hold near the hospital, waiting for the weather to clear. Additionally, I wanted extra fuel to go back to the alternate hospital or hold nearby if the storm near the scene LZ delayed initial landing. Fuel = options.

Since the patient report lacked detail and, statistically, we only fly about 30% of the Alert-1 calls received, I decided to wait for more facts before pumping an additional 30 gallons. If the call canceled, or the patient had trauma, the additional fuel would not be necessary and would seriously reduce the useful load we could fly from the base on a subsequent call. De-fuelling would require nearly an hour's flight to burn off at significant cost. Just in case, I readied the fuel truck and laid the hose next to the helicopter to expedite the fuelling process. As I asked the medics to request a first responder ETA, someone on the net requested that we pre-launch. I repeated the request for their ETA and was told about four minutes. Since the fuel decision was important, if not critical, I elected to wait for first responder arrival before launching.

My priorities are crew safety, patient welfare, customer support and cost in that order. Transporting the patient to the preferred hospital outweighed the negatives of a short delay. I felt that proceeding to the intended destination without contingency fuel was imprudent, unwise, compromised safety and was possibly in violation of 135.69, and the [Operating Manual]. Based on my considered experience, I thought my decision was the best balance of safety concerns, regulatory requirements, patient outcome, and costs.

When the word came in that the passenger was a burn only, I quickly fuelled the helicopter and launched within a few minutes. After the passenger was loaded, we proceeded south and watched the thunderstorms dissipate on the radar allowing us to get into Hospital [X]. The mission was completed without further incident.

AFTERMATH: The Base Coordinator asked that [I] respond to an inquiry from the EMS Chief who was concerned about the fact that I had delayed a launch after being told to do so.

I responded that the customer merely makes transportation requests, they do not have the ultimate authority to order a launch. That authority rests solely with the PIC under FAR 135.67 and the [Operating Manual]. I explained my decision, informed him that I stood by my decision and told him that the Chief was welcome to discuss it with me if he had any further concerns or questions. I was later told that the Chief disagreed with my assessment. I expressed my displeasure, informing the coordinator that neither the Chief, [X] County Dispatch or anyone other than the Certificate Holder, may exercise operational control over the [flight] program. (See FAR 119.9, 119.43, 135.79 and Operating Manual) To wit: "Only a PIC who is a direct employee of [Corporation] may exercise this second tier Operational Control over any [Corporation] flight." "At no time shall any non-certified entity attempt to exercise Operational Control nor hinder in any way, [Corporation] oversight and/or exercising of Operational Control..." "The...call center (County Dispatch or Command) has no authority to override the authority of [Corporation], or the pilot's authority to refuse any mission, or other flight safety issues. At no time during a response to a medical flight will speed into action criteria be allowed to compromise safety. Failure to adhere to the certificate holder's directions and instructions may be subject to legal enforcement action by the FAA."

My actions also complied with FAR 135.23, 135.25, 135.65,

135.71, 135.209. [Corporation] Risk Assessment Program was also utilized. Operating Manual Condition for patient leg was indicated to be "C." [Operating Manual Requests for Service states: "No attempts should be made by the Communications Specialist (in this case County Command) to question the pilot's judgment in denying a flight request."

Lessons Learned:

Since I am now being *questioned* about my decision(s) by the customer and my supervisors contrary to the [Operating Manual], the issue of operational control needs to be addressed.

If this isn't an operational control issue, then I don't know what is. Being questioned and second-guessed about conservative, pro-safety PIC decisions constitutes pressure to fly. Therefore it is a serious (indeed major) safety of flight issue which needs to be addressed at the highest level of management.

NOTES:

LOOSE ARTICLES LOST
CALLBACK

AS332, Government, Ferry, Phase: Takeoff, Aug 2011

Upon departure from the airport I was repositioning my kneeboard from the door pocket to my knee. As I brought the kneeboard up and away from the door pocket it passed within a few inches of the open pilot door window.

It was immediately sucked from my grasp and departed the aircraft. A person in the cabin witnessed the kneeboard strike the port side sponson and then [it] continued down and away from the ship.

The sponson received minor damage.

We normally fly with the rear doors closed on this ship. However, due to the extreme heat, we chose to fly with the doors open this day.

Lessons Learned:

In this configuration, there is a significant flow of air forward which contributed to the negative pressure in the cockpit.

NOTES:

RESTRICTED VIEWING FIELD
CALLBACK

EC135, Commercial, Phase: Parked, Name Withheld, May 2015

I am submitting this report to highlight a safety issue that is present with certain EC135 NVG Cockpit Lighting STCs.

The safety issue is that when chin bubble floor mats are utilized while flying NVGs, the pilot's viewing field is restricted / limited while conducting landing operations in our flight profile. My primary concern for submitting this report is to express that the use of chin bubble floor mats, to comply with the STC, is an extremely unsafe practice. Glare shield curtains should be the Company's standard when given that option within the STC.

Some NVG cockpit STCs require either the chin bubble floor mats or the cockpit glare shield curtains to be installed for NVG operations. I operate one of two EC135s as our primary aircraft and from time to time we operate another as a spare. I have found differences in what is or is not required by the STC's for each of these aircraft.

One STC requires that either the chin bubble floor mats or the cockpit glare shield curtains be installed while using NVGs. While another STC does not require either in the limitations section, it only has a note stating that if external lights and/or cultural lighting causes distracting glare and reflections, the use of chin bubble carpets is recommended.

All of these aircraft have similar landing light configurations yet only one makes it a requirement, a limitation, for either to be installed for NVG operations.

I brought this topic up at a base safety meeting to see if the cockpit glare-shield curtains could be installed in the one aircraft that requires either. Our mechanics said that the glare-shield curtains could be installed and are currently waiting for upcoming scheduled maintenance to install them in the aircraft that has the requirement.

Once again; utilization of the chin bubble floor mats restrict / limit the pilot's viewing field while conducting NVG operations which is an extremely unsafe practice. Glare-shield curtains should be the Company's standard when given that option within the STC.

Lessons Learned:

The reporter stated that two types of Night Vision Goggle (NVG) instrument light filters are authorized for the EC-135 by an STC. One is a curtain which snaps in on both sides of the cockpit then attach with Velcro along the glare-shield front edge to hang down blocking extraneously light from the pilot's instrument. Pilots can see down through the chin bubble for obstructions during steep descents. The other type of instrument light filter is a black mat which lays over the chin bubble blocking out upward light which may interfere with a pilot's NVG instrument vision.

The reporter does not believe the blackout matt is safe because all vision down through the chin bubble is blocked restricting the pilots' vision of obstacles and making landing descents less safe.

NOTES:

NGV WITHOUT CERTIFICATION
CALLBACK

Bell 222, Ambulance, Phase: Initial Approach, Dec 2011

While flying on an inter-facility flight in a NON NVG [Night Vision Goggles] modified aircraft, I had NVGs mounted on my helmet and used them momentarily to ensure terrain clearance and to identify known obstacles on my approach paths. I did not fly an NVG profile or use them in the entirety of the flight. I used them only momentarily to ensure terrain clearance and to quickly ID obstacles on my approach path. 99% of the profile was unaided with my using standard night flying techniques such as appropriate altitudes, known land marks, and supplemental lighting (Searchlight & Night Sun).

To ensure a safe flight route, I maintained the appropriate altitudes until I arrived at known landmarks that ensured terrain clearance prior to any descents and used outside visual references with a good scan without using the NVGs. I used them strictly momentarily and my intent was to enhance safety and not to flagrantly violate any regulations, instructions or directives. If I did so it was purely unintentional.

I used poor judgment and should not have utilized the NVG's in an un-modified aircraft. I accept responsibility for my actions and realize I acted incorrectly. While my intentions were safety related, my actions were incorrect.

Lessons Learned:

I will ensure this will not occur again by not taking NVGs on board a non-NVG modified aircraft and flying unaided when conditions are appropriate and not accepting the flight if the conditions are not appropriate for unaided flight.

<u>NOTES:</u>

OPERATIONAL CHECK FLIGHT
CALLBACK

Bell 206, Ambulance, Phase: Parked, Name Withheld, Feb 2010

I am employed by an air ambulance service based out of ZZZ. On the date listed, the company's Bell 206 L1 aircraft, was taken out of service around XX:oo hrs the previous date for some required maintenance. One of these required maintenance items was the replacement of a hydraulic servo which was due for overhaul.

At approximately XY:45 hrs the Mechanic had approached me and informed myself that an engine run was required in order to check for any leaks of the replaced hydraulic servo.

As per his request this engine run was completed at 100% power and the controls were manipulated to check for any type of binding and or abnormalities. None were noted and the aircraft was returned to the hangar pending completion of the required maintenance.

I was later approached by the Mechanic where I verified that the required maintenance was properly documented and signed off in the engineering log. Upon inquiring if a test flight was required, I was informed by the Mechanic, that per his documentation, no test flight was required and verification that a leak check was performed was all in which was needed. Based upon this I returned the aircraft back into service, and the mechanic left the base upon completion of faxing in his required documents.

It was a short time later we were requested to take a patient flight from an outlying hospital into the ZZZ1 a more advanced hospital. Upon returning back to base, a Mechanic Supervisor checked the Mechanic's paperwork which had completed the maintenance on the aircraft and noted that no operational check flight had been performed prior to returning the aircraft to service.

I checked the company operational manual per the requirement of operational check flights, and noted when the required checks are to be performed. Per the manual, an operational check flight is performed for a number of reasons, one of these being "wing, horizontal stabilizer, vertical stabilizer, or helicopter primary flight control change. This does not apply to the removal and reconnection of flight control linkages unless the length has been adjusted. Note: primary control is the elevator, collective and cyclic controls, and anti-torque controls."

Upon reviewing this, I felt that a flight control linkage can be removed and placed back on without the requirement of a test flight, and being there is no adjustment made for a hydraulic servo, that this same no need for an operational check flight would have been required.

Lessons Learned:

This was later viewed by the administration with the company and determined that although a bit vague and could lead to misinterpretation by others, that a required operational check flight was mandatory.

Due to this, there is a possibility of later changes being made to the operational manual reference to the operational check flight requirements — these operational check flight requirements being found under Revision XX of our company's maintenance and servicing and operational check flights.

This aircraft was shortly thereafter flown on an operational check flight, and same was noted within the aircraft engineering log.

NOTES:

FUMES MISIDENTIFIED AS SMOKE
CALLBACK

S-76, Passenger, Phase: Takeoff, Name Withheld, Jan 2011

Prior to takeoff I had turned on the cabin heater due to the low outside air temperature. Upon applying power and lifting off the elevated heliport to the north, the Flight Medic in the cabin area stated she smelled something "funny." As I turned westbound she said, "I've got smoke in the back." I asked her to confirm that she had said smoke and she replied, "Yes and it is pretty thick." I was already in contact with ZZZ Tower as we were exiting the Class B airspace. I asked the Tower for permission to land at the FBO ramp area on the north side of runways XXL and R. I stated, "I have smoke in the cabin." The Tower responded that I was cleared to land abeam the ramp on the taxiway. She also asked if I needed assistance. I stated, "Not at this time." She asked me if I wanted the emergency equipment to roll and I replied again, "Not at this time."

I complied with the immediate action items on the aircraft checklist which included opening the pilot's door window, turning off the environmental control unit (heater system), and closing the ECU vents. As soon as I did these items, the flight medic said the smoke cleared. Opening the pilot's vent window hastened the dissipation of the smoke and fumes. Upon landing and taxiing to the ramp, we opened the cabin and cockpit doors to further ventilate the cabin and cockpit area.

I determined that the smoke was actually the residual water and cleaning solution which had been used earlier that day after a morning flight when the engines were rinsed due to flight in a salt water environment. The fluid had not evaporated and was trapped in the bleed air lines that are part of the heater system. When the heater was activated, the water and cleaning fluid vaporized, causing a fogging of the cabin that resembled smoke in the darkness of the rear cabin. At no time was there any actual smoke. I recognized the smell for what it was but was determined not to make any further diagnosis in the air.

Once the smell and vapor cloud had cleared, I was confident that the problem had been with the heater system and that there was no danger of continued flight. We then departed and returned to our home base. My diagnosis was confirmed by maintenance personnel and the daytime duty pilot when they performed a ground check of the aircraft later that morning. They saw only some residual vapor and smelled a slight chemical odor from the heater vents.

Lessons Learned:

The Mechanic confirmed my evaluation of the chain of events and recommended running the aircraft engines after any engine rinse to ensure all residual fluid was eliminated from the bleed air lines to the heater.

<u>NOTES:</u>

I'VE HAD AN ACCIDENT
CASA

R44, Commercial, Phase: Emergency Procedure, Dayle Jordan, Oct 2003

I'm the manager of Cape York Helicopters, a Cairns-based helicopter charter company with four Robinson R44s, a base in the Torres Strait and five pilots on the books. It's a busy job, and the phone rings all day, with lots of calls from our pilots. Much of our charter work is in very remote places, and we often hold search and rescue (SAR) information for our pilots.

Even if we're not holding their SAR, our pilots contact us frequently throughout the day, which is for ease of backtracking (if we ever do lose contact), operational convenience, and peace of mind on the big days. The helicopters are all fitted with satellite phones and we also get limited CDMA coverage, so this frequent contact is easy and just part of the way we do things. February 12, 2003 was just like any other day. It had only been a couple of weeks since we moved into our new office, and Yvonne Wallace (director of the business), was planning on getting away early to pick up new blinds for the windows.

We had two choppers in the air: Leon was flying around the Torres Strait and Paul was north of Cairns with Telstra. Both pilots had called in a couple of times already. The phone rang at 1.30pm: "Dayle, it's Paul, I've had an accident". Yvonne was on her way out and I grabbed her arm as she walked past me and scribbled "ACCIDENT!" on the back of the first piece of paper I laid my hands on. As I asked Paul questions, I kept scribbling:

"3nm nth mba airport massive vibration hit powerlines everyone's OK." Paul could see some people coming towards him, so I told him to make sure everyone was safe, to sit down till help arrived and that I'd call him back. Yvonne took off for Mareeba, to be on the scene. That's when I reached for the emergency response plan.

It used to live on a hook within reach of my chair in the old office, and it had been hanging there for more than two years without ever being used, but I'd dusted it off and carted it up to the new office with me. It was just inside the cupboard on my right, and within seconds it was in front of me. Our emergency response plan is loosely based on one I was familiar with in the mining industry, and consists of two pieces of paper – a flowchart and an accident notification form.

Following the flowchart, the first thing to do after receiving news of the accident is to appoint an emergency response coordinator (I appointed myself), who should complete the accident notification form. On the form you record the date and time of the accident, the name and contact numbers of the person who reported the accident, what was said, the names of the pilot and passengers, the nature of any injuries, and the time and details of all calls made after the initial notification. (When we reviewed the form, we made that last section a lot bigger.)

The flowchart ensures you contact police, ambulance, fire, ATSB, energy and phone companies if necessary, the client, the pilot's family, insurance company, maintenance organisation, bank manager, other company bases and staff.

It also suggests you prepare a press statement, assuming you will be battered with press enquiries, though we found that some of the press just made up the story without even contacting us!

Lessons Learned:

On the day, the plan was a rock to cling to. During the chaos of all the notifications and the followup calls and then everyone phoning as they hear about the accident, not to mention all the calls not relating to the accident, it would have been easy to forget things.

I was dealing with my emotions: my pilot was hurt and I had to tell his mother. I was also wondering where we were going to get another helicopter, and how much business we would lose. Following the plan allowed me to go with the flow and know that it was all getting done.

Completing the notification form, especially recording all the calls, was also a huge comfort, because I could check who'd called, what time it was, what they'd said, what I'd told them.

We reviewed our emergency response procedures within a week of this accident, and spoke to the emergency helicopter rescue team in Cairns, and the Australian Search and Rescue (AusSAR) guys in Canberra. The following improvements were introduced:

- Ask the pilot specific questions about injuries, remembering he or she may be in shock. In this case, Paul said he was OK, when in fact he had some pretty serious injuries.
- When you employ a pilot, and you ask for an emergency contact number, ask for half a dozen of them. We had Paul's mother's home number, but not her work number nor her mobile number, and we couldn't get in touch with her for some hours after the accident.
- If the pilot or passengers are being taken to hospital, arrange for someone to meet them. Although it wasn't on the plan, we did think to do this on the day, and saved

Paul from having cameras and microphones shoved into his face as they were pulling him out of the emergency chopper.

- Prepare a document that shows the location of switches and power sources on your aircraft. Following an accident, it should be faxed to the police who will attend the scene. (They will probably have no idea how to stop fuel flow or turn off power sources, yet they will command the scene.)
- Get statements from witnesses, pilots and passengers as soon as practical, while details are still fresh.
- Lodge a flight plan or leave a flight note with a responsible person.
- Get an emergency contact for passengers when they book the flight. And remember, a work number is of little use if the accident happens after hours. AusSAR also recommends recording passengers' medical conditions at the time of booking to assist with medical treatment.

Now that we've lived through our worst nightmare, here are a few tips and tricks that might help other operators in the future:

- Have an emergency response plan. It doesn't have to be complicated, but make sure everyone knows where it is and how it works, and review it periodically.
- Get a press statement out to the media, even if they don't ask for it. In this case, the media reported the accident as "helicopter hits power lines", which is vastly different from "helicopter experiences massive vibration at 1000ft and hits power lines during uncontrollable descent".
- Be up front with your clients. We phoned our main clients within days of the accident to clear up the "helicopter hits power lines" story, and as soon as we knew what CASA and the ATSB were doing about the

investigation, we wrote to our clients and told them. After
the ATSB released its findings, prompting another splash
in the press, we wrote to our clients again, enclosing a
copy of the report.

- Above all, always do the right thing, and be in control of
all the controllable elements of this high-risk business.
When the unexpected happens, it's your day-to-day
procedures and your safety culture that supports you.

Dayle Jordan is willing to share Cape York Helicopters' emergency response plan with other operators. She can be contacted at info@capeyorkhelicopters.com or phone +61 (07) 4093 0250

ANALYSIS:

Fortunately, this company had an emergency response plan in
place and, as a result, coped admirably with the aftermath of this
accident, despite the obvious difficulties.

Emergency response planning is all about mitigating the consequences of accidents An effective plan can save lives, minimise
injuries and reduce damage to property and the environment. The
way you respond in the hours after an accident may also determine
the survival of your business. While this company responded well in
the circumstances, its lack of familiarity with their emergency
response plan no doubt added to the stress of the day.

Unfortunately, it is not uncommon for emergency response plans
to be dusted off and used for the first time in a real emergency. Periodic simulations and "roundtable" exercises ensure key personnel are
familiar with their roles, while allowing staff and management to
identify and address shortcomings in the system. The Flight Safety
Foundation says: "To have an accident is unfortunate, to have an accident and learn nothing from it is unforgivable."

To its credit, the company reviewed and improved its emergency response procedures soon after the accident. Its internal communications throughout the experience were also commendable. By installing satellite phones in its helicopters and talking to its pilots throughout the day, it ensured it would be quickly apprised of problems. Consequently, it was among the first to find out about the accident, and was able to respond quickly. If the accident had occurred in a more remote location, this may have had an even more significant bearing on the outcome.

If you are thinking about developing an emergency response plan, there is a few things to consider. The plan must:

- Be relevant and useful to the people on duty at the time of the event;
- Include checklists and emergency contact details;
- Be reviewed and updated regularly; and
- Be simulated to ensure the adequacy of the plan and the readiness of people allocated key roles and responsibilities.

Emergency response planning should be an integral part of every organisation's safety management system. An effective plan may not only save the lives of your crew and passengers, it may also save your company.

Jo De Landre
Human Factors Specialist

Jo has worked for the ATSB and CASA and is currently the principal human factors consultant with *Safety Wise Solutions*.

NOTES:

NOT ENOUGH TORQUE
CALLBACK

EC135, Passenger, Phase: Parked, Name Withheld, May 2013

Aircraft X, a Eurocopter EC-135 was placed into Maintenance for its 800-Hour inspections. While complying with the aircraft inspections, the A-2411 Datasheet [indicated] a 1600-Hour inspection was due. The main rotor hub and the main rotor hub shaft were removed in accordance with EC-135 Maintenance Manual Chapter 62-31-00 (4-3), Removal Main Rotor Hub-Shaft, with no incident. We then carried out the inspection of the spacer tube. Upon reassembly of main rotor hub-shaft in accordance with EC-135 MM Chapter 62-31-00 (4-4), the bolts that attach the hinged support to the bearing block just aft of the mixing unit of this aircraft were installed. While installing the main rotor hub, I was on the lift that was used to assist with the install. Mechanic Y was in the aircraft watching the shaft to make sure that the splines were inserted into the grooves correctly. Lead Mechanic X was on top of the aircraft watching the shaft from the top of the aircraft; it was at that point that I believe that the bolts were installed. These two bolts were not torqued or cotter pinned. The aircraft was then moved back into to the hangar and reassembly resumed. This was not noticed at time of reassembly of aircraft for maintenance flight following 800-Hour inspections.

The aircraft then flew its maintenance flight which lasted about 2-hours total flight time. The aircraft returned to ZZZ to finish up paperwork and to be placed back in service in ZZZ1. The pilot came down to swap Aircraft X from Aircraft Y, which they were currently in so that the pilot could take Aircraft X back to ZZZ1. Aircraft X was [later] placed back into service at the ZZZ1 Base. I wanted to make sure my night pilot was good with all the paper and to see if he had any questions over anything that had been completed during the maintenance evolution. He informed me that the day pilot stated that there was a 1 to 1 hop in all axis of flight; I then proceeded to ask him to let me know what he felt about the aircraft after he flew it to see if he could duplicate the 1 to 1 hop in all axis of flight.

I arrived to work this morning, upon my arrival Aircraft X was just starting to start engines on a patient flight. They completed patient flight without an incident. I walked out to talk to the pilot about 1 to 1 hop in all axis of flight and he said that it was there and seemed to be getting worse. I then stated to him to place Aircraft X on a 30-minute maintenance delay and for us to take the aircraft for a flight to see if I could identify the 1 to 1 hop in all axis of flight. Aircraft X was then placed on a 30-minute maintenance delay. Before we got into the aircraft the pilot and I took a look at everything from the hub cap down to the mixing unit. That's when I noticed something was not correct with the hinged support attachment to the bearing block. I then pulled the Co-Pilot's side transmission cowling and noticed the bolts securing the hinged support to the bearing block on the Co-Pilot's side were not torqued nor cotter pinned. I then stopped everything took Aircraft X out of service and made a notification to Lead Mechanic X. After speaking with Lead Mechanic X, I was directed to call the Area Maintenance Manager who advised me to take the other transmission cowling off to inspect the other bolt and to take pictures and send them to him.

I removed the other cowling and it was noticed the bolt was

installed and the nut had worked its way off the bolt. I did a search for the nut and it was found in between the main transmission and the mixing unit. Once the nut was located I took pictures from every angle that I could see and then sent them to Maintenance Manager and Mr. X. I informed the Maintenance Manager, Mr. X and my Lead Mechanic of where the nut was located. While informing Mr. X of where nut was located, he directed me to call American Euro-copter Technical Support Representative, Mr. Y who was out to lunch at the time I called. Mr. Y then returned my phone call and I informed him of what was going on and he advised me to send him information stating what had happened and how it was found, along with pictures of the area. He also advised me that he would forward the pictures and statement to the design department for review and that we should have an answer back in the morning to advise us what was needed to place Aircraft X back into service.

Lessons Learned:

I believe that there is certain action that could have been handled differently. We could have had another Mechanic that was not there during the maintenance, come in and inspect everything before that aircraft was placed back into service.

The extra set of eyes from someone that was not there at the time of install would have been very beneficial in the aid of identifying issues before the aircraft was placed back into a state to which it was ready to be flown for the post 800-Hour flight.

I also believe another Mechanic could have been on hand to assist with the installation of the main rotor hub shaft.

NOTES:

ENGINE INSTALLATION
CALLBACK

EC135, Phase: Unscheduled Maintenance, Name Withheld, Feb 2017

Inspector:

A maintenance technician and I began packing an engine for shipment and found the pin to the aft mount would not fit. We discovered bushings that needed to be removed in order for the pin to fit and, I instantly wondered if the new engine we just installed had the bushings installed in the aft mount.

We originally removed the #1 engine to facilitate Maintenance. The original plan was for a Pratt and Whitney Team to swap a combustion liner in the field and this engine would be reinstalled. After removal, I performed an inspection of the engine mounts and all mounts, fittings, and bushings checked out ok. The Pratt and Whitney team arrived and during their combustion liner swap discovered more damage and the engine would need to go to the manufacturer for repairs.

The decision was made to replace the engine. The engine arrived, [the maintenance technician] and I began installation [and] engine installation went well we continued installation and completed installation on IAW EC135 MM. During check flights we discovered Maintenance Manuals on company Web site was not current. We notified [the company] and was informed that current revision will be installed. We completed Check flights and signed off the aircraft return to service.

The aircraft logged around 2 hours since engine replacement. The aircraft arrived this afternoon and [another maintenance technician] and I checked the aft mount to discover bushings not installed. We took the aircraft out of service and made repairs. We inspected all handwork and did not find any discrepancies.

We conducted all maintenance according to Maintenance Manual (MM) and I am still disappointed in myself for missing those bushings. I believe the error came from us installing what we believed to be a complete and serviceable engine. Minor build up was required and manuals where used during all maintenance. Because a late decision was made to install a new engine I did not think to inspect new engine mounts after all mounts had already been inspected. The bushings are tiny and hidden in the aft engine mount and easy to miss I did not think to check aft mount until we had to remove them for engine packing. We did work late hours during this whole event but nothing unreasonable at no time did I feel too tired or unsafe.

[The maintenance technician] and I reviewed the RII and did not see mounts. One remedy to this problem I would like to recommend is to add a caution or note in the RII that includes special attention to the aft engine mounts for bushings.

Technician:

The original plan was to remove #1 Engine from airframe to have [manufacturer] perform on-site replacement of Combustion Liner. Upon vendor completion, same engine was to be re-installed on the airframe. EC-135 Inspection of Engine Mounts was performed at this time. During vendor inspection, it was determined additional internal depot-level repairs were recommended, and another engine was ordered.

Engine was originally scheduled for delivery by XA30 Local Time. Plan was to install engine with our Lead Mechanic as [another technician] was out of town on company business. The engine did not show up as scheduled, and was delivered 24 hours later. Due to this fact, a decision was made for the mechanics to come in to install the engine. I was not feeling great, and had requested to perform Maintenance on [another day]. The decision was made to continue with engine installation [on a later day].

[Another technician] & I arrived and began engine build-up, following the manuals printed that morning. The new engine was mostly complete, however some small components such as the Air Box Enclosure, Water Injection Fitting, HMU Manual Control Linkage , etc., - still needed to be swapped over from the previous engine, still on the Engine Stand provided by the P&W Vendor. We had the engine physically installed by that evening, however we both began to feel we needed to stop for the evening as it had been a long day, we had made good progress, and we were reaching the 12-hour mark. Additionally, the hangar lighting is not very bright and there are no provisions for additional lighting as the entire hangar only has a single point, 4-Outlet Electrical Outlet. We decided to make notes of where we left off, clean up, account for tools, and re-inspect everything on [following workday] Morning, which we did.

[Later], we looked over the installation, and verified all bolts and lines were torqued and safetied. We continued on with Driveshaft Installation, took another good look together, and proceeded to install firewalls, ejector, cowlings, etc. We opted to wait to complete Leak Checks, and Operational Flight Checks due to inclement weather outside to ensure a proper leak check. We [then] performed leak checks and installed cowlings in preparation for ops flight checks. We flew for approximately an hour following the procedures printed that was supposed to be up to date, and encountered an issue with our high altitude N_2 adjustment. We returned to base and contacted a tech rep who informed us our manuals were out of revision. We accessed and printed the current manuals.

An email was sent to inform her of the disparity, and that it created a delay in the return to service. This would normally only be an inconvenience however, we needed to complete the maintenance to free up the spare aircraft for use at another base as they were awaiting parts that were scheduled for delivery for an inspection that expired a [following date]. We resumed flight checks with the current manuals and the flight checks were completed without issue. At no time during the flight did we notice any abnormal operating conditions such as vibration or noises resulting from installation. The aircraft was subsequently returned to service.

[On another date] I arrived as the aircraft was departing for EMS support while their base aircraft was undergoing maintenance. [Another technician] arrived shortly after. [The other technician] & I began working on removal of old engine from the Engine Stand and attachment to shipping crate in preparation for return to manufacturer for repairs. During installation of the mounts on to the engine, I discovered the Clevis Pin did not fit through the Aft Engine Support Boss. After looking at it for a moment, we revealed the cause to be due to bushings installed in the boss. We removed the bushings and inspected them, and began to wonder if the new engines came with bushings or if they perhaps had been forgotten. The aircraft arrived back shortly thereafter.

Upon landing, [another technician] and I greeted the aircraft and spoke to the pilot; He reported no anomalies with the aircraft. We inspected the R/H Engine & compared to the L/H Engine; upon which no determination could be made. After retrieving a 1 Inch Inspection Mirror, we found the L/H Engine (Newly Installed) to be missing the Bushings from the Aft Mount (Z-Strut). We immediately documented the discrepancy in the Aircraft Logbook, informed the pilot, and notified [the] Lead & Regional Maintenance Manager the Aircraft was out of service due to our findings. We printed current manuals, gained access to the #1 Engine "Z-Strut", and supported the

Engine with a hoist. Once completed, we each verified no tools, rags, or foreign debris were present, all lines and fittings were torque and safetied, and requested the pilot to be a third set of eyes as well. Upon sign-off of maintenance performed and verification all tools were accounted for, an Operational Check.

Flight was accomplished to verify satisfactory Installation and the aircraft was returned to service.

Lessons Learned:

After the aircraft was returned to service, [another technician] and I reviewed the RII Checklist again to see how this could have been missed. We agreed it needs revising.

The closest the RII Checklist comes to the mounts is where it states "Thoroughly inspect all maintenance areas for proper installation of Safety Wire, Cotter Pins and Lock Tabs. In limited sight areas use a flashlight and mirror if necessary."

NOTES:

CHAPTER 5
WEATHER

"One day I'd been forced down in bad weather at a country airstrip and a helicopter came in and landed. I walked over to the pilot and I said this helicopter, is it designed to fly in cloud? He said no, you just fly under the cloud. And if ever the cloud gets too low, you land and have a cup of tea with somebody. Wow I thought, so I ordered a helicopter, learnt to fly it and the rest is history."

Dick Smith, Helicopter Pilot and Adventurer

FROM VFR TO IMC
CALLBACK

EC135, Commercial, Flight Phase: Cruise, Name Withheld, July 2017

I was sent to the maintenance facility to pick up a spare aircraft and conduct a ferry flight.

During my pre-flight planning and weather analysis, I noticed that the METAR at my destination [airport] was reporting MVFR conditions with the TAF forecasting continued MVFR conditions for my expected arrival time. I also noticed pockets of MVFR ceilings along my route of flight.

Since I had not received my 297/IFR check yet and I'm a VFR only pilot for now, I decided to file and activate a VFR flight plan for the ferry flight.

En route, I encountered a stable deck of MVFR ceilings but the ceilings seemed lower than the 1,500' AGL OVC being reported. There were also scattered clouds at or below 1,000' AGL that I was having to side step and avoid. In addition to the low ceilings, I also encountered several towers that were hard to see in the low light and several crop duster planes doing low-level work at or below 500' AGL estimated.

The crop dusters were not communicating on any of the advisory frequencies and I saw one fly underneath me at 200' AGL. I attempted to steer away from the low-flying crop duster but the pilot did a climbing, looping turn and was flying towards me again. I had visual contact with several other low-flying crop dusters but none of the others got as close to me as the first one.

As the ceilings continued to drop and push me closer to the ground, I decided to contact TRACON for VFR flight following and radar services. Coms with Approach were weak but they were able to pick me up on radar. I contacted FSS to inquire about the weather at my destination and was told that [the destination] was now reporting IFR and 900' CIG. By that point, I decided to request an IFR pop-up with Approach. Although I have not had my 297 check yet, I feel comfortable flying IFR and maintain my IFR currency flying in the [military].

My decision-making to file IFR en route was in the interest of safety and was impacted by the following contributing factors:

1. The ceilings en route seemed to be lower than reported.
2. I encountered several crop dusters operating low-level in Class G and not communicating with one flying underneath me.
3. Some of the towers along my route in the low light and OVC cloud cover.
4. Weather at my destination had fallen to IFR conditions.
5. I have practiced the [the approach] under VFR conditions and felt comfortable with the IAP.
6. At my low altitudes AGL for cloud clearance, I was starting to lose coms with ATC for VFR flight following. I was too far from my home base to talk with my base OPS and I did not feel comfortable not having flight follow coms while "scud running" over unfamiliar airspace.

7. I was conducting a ferry flight with no Passengers under Part 91.

Lessons Learned:

Wait for weather to improve along my route of flight and not accept ferry flight missions in the future when there is a high possibility that I may have to request IFR to accomplish the flight.

NOTES:

FLYING IN THE MUCK
CALLBACK

EC130, Ambulance, Phase: Cruise, Name Withheld, May 2017

The crew and I just completed a scene flight and were at the hospital. I refuelled the aircraft and double checked weather prior to making our return leg to base.

All weather reporting stations along our route were all reporting VFR. The closest weather reporting station to our base, was reporting winds 240 at 8 knots, 10 SM visibility, ceilings 9,000 feet scattered, temperature 75 degrees, dew point 64 degrees, humidity 84%, and about 16% illumination. Winds at 1,000 feet AGL was 250 26-28 knots.

We departed to the northeast enroute back to base at 2,000 feet MSL, which is a 20 minute flight.

Approximately 15 minutes into the flight I noticed some small patches of fog below us at about 500 feet AGL. Our route takes us along the river, and we were approaching a power plant with several bright lights. Once we got to the power plant, the crew and I realized it was very hazy and I decided to deviate from our route and head northbound toward a local Airport. I could still see ground lights and cars driving on the roads. Weather wasn't looking much better to the north, so I made a slight left turn toward the northwest to attempt to get away from the river. I also elected to start a climb in the event we went in Inadvertent Instrument Meteorological Conditions (IIMC).

I referenced the GPS and we were 11.3 NM south of the airport now at 2,500 feet. I already had the UNICOM frequency tuned in so I attempted to activate the airport lighting via radio clicks. I saw no signs of lights to the north, and ground lights were deteriorating directly below us. I told the crew we were IIMC and I was coming inside and committing to instruments.

I followed the IIMC procedure and got established on a north-west heading. I knew we were in the clouds at this point because the strobe light on the belly of the aircraft was reflecting into the cockpit, so I turned it off. Once I was at my MSA of 4,000 feet MSL, I made a small left turn to 270 and planned on recovering to a county airport as we had just been there previously and I knew it was VMC there when we left about an hour prior. I then switched up Approach control, which was in the standby frequency, and established communication and let them know I was [requesting priority handling] for IIMC. I elected to maintain our company discrete squawk code. I stated my altitude and heading and requested radar vectors to the County Airport.

Once they had me on radar he had me turn left to 220 and that would put me on a track toward the airport, which was 14 miles away. After flying this track approximately 5 minutes we broke out of IMC conditions. I told ATC that I was now VMC and had the Airport in sight, but was going to stay committed onto the instruments until I got closer. I told ATC I requested to get set up for the ILS into the county airport for planning purposes. I then descend down to glide slope intercept altitude of 2,500 feet as I was still currently VMC and wanted to make sure I could stay that way at a lower altitude. Approach then advised me that they knew I said I had the airport in sight, but wanted to confirm and the Airport was 12 o'clock and 5 miles. I stated that indeed I had the airport in sight and could cancel the clearance and descend down to the airport VFR. We landed at the airport with no other issues.

Lessons Learned:

Not much we can do to remedy this issue, except add more accurate weather reporting stations, especially in known troubled areas.

Took Evasive Action, Requested ATC Assistance/Clarification, Diverted, Became Reoriented, Air Traffic Control Issued New Clearance.

<u>NOTES:</u>

EMERGENCY DECLARED
CALLBACK

BHT-407, Ambulance, Phase: Cruise, Name Withheld, Mar 2016

When I arrived, weather looked to be clearing no ceiling and 10 miles visibility. I checked the forecast it was almost 3 hours old. I checked different locations for the trending weather. It looked good for a 10-15 min flight down south to the sending hospital.

On the flight to the sending hospital the weather appeared to be moving to the north. On the return flight the weather became visible at about 5 miles south of town and was moving west to east. I checked the ATIS it was 400 feet scattered 1,600 feet overcast. I told the crew that we would have to land at the airport instead of the hospital. At that time I could still see the airport. Told the tower that I needed to land at the airport. The tower then stated that the weather was now 600 feet overcast and I was losing sight of the airport. The weather was moving quickly so I decided to get vectors to the approach instead of looking for VMC and landing. The tower asked if I wanted to coordinate with center I said yes. Then I started the check list for Inadvertent Instrument Meteorological Conditions (IIMC) to get my wings level, heading, adjusting climb power and airspeed.

Then the tower called and said to contact Center.

So I switched and called Center, he immediately gave me a code, heading and altitude which I repeated and started the request.

This is where I was "distracted".

I started setting the radios getting the approach plate out and I should have "declared an emergency" at this point and I didn't. I was talking to the crew explaining what was going on told them that I might need them at some point and to be available.

Center vectored me to the final approach fix and I did the approach.

I broke out of the clouds at 600 feet AGL and 2 miles from the runway. I air taxied to parking where the ambulance was waiting for the patient. When the crew returned I did a post flight brief asked them if they had any concerns or questions. They said that I kept them informed on what was going on and had no complaints or questions.

Lessons Learned:

I think what happened was I was close to the airport and already talking to the tower he was trying to help. So he got Center all my information and when he passed me off to center I was busy starting my IIMC procedure wings level and establishing a climb.

When I contacted Center he immediately gave me the directions and I just complied, this is where I should have declared the emergency.

We practice this procedure and I know what I should do I just feel I got distracted. I was flying the aircraft setting the frequencies getting out the approach plate and informing the crew. I feel this is where the workload may have been part of my distraction.

The IIMC procedure worked very well, I failed to get back to the check list and declare that it was an emergency because we are not certified for IMC. I feel that when I got distracted I got off the check list and that caused my error.

I will not let it happen again I will be more diligent with my check list.

NOTES:

UNUSUAL ATTITUDES
CALLBACK

Bell 206 LR, Passenger, Phase: Initial Approach, Jan 2011

The mission was to transport two offshore platform operators from the airport to the platform and pick up two operators to transport back to the airport, normally a 25 minute one way flight. A cold front trough was moving into the area.

Pilot checked TAF, METAR, and radar NEXRAD images from National Weather Service to ensure sufficient time would be available to make the round trip and after embarking the passengers we took off. The outbound flight took about 5 minutes longer than usual and the helicopter also visited a satellite platform at operator request, which took an additional 9 minutes. After embarking the two returning personnel the helicopter departed the offshore platform for the airport, cruising between 400 and 1,000 ft MSL.

Approximately 18 miles south southeast of the airport visibility started to decrease due to rain and fog obscuration of the western shore, the intended route to the airport. Pilot had already obtained ATIS and contacted the airport Tower to check for any inbound traffic and then Approach for a special VFR clearance, which was issued with a transponder code.

While this was happening pilot-in-command was turning further east of course experiencing instinctive reluctance to turn towards the approaching weather and encountering increasing turbulence and rain while maintaining visual contact with the surface of the bay.

Next the aircraft encountered some intense rain and an updraft that took the helicopter to 3,000 ft MSL. Visual contact with surface was lost. During this sequence of events aircraft angle of bank reached 30 degrees right and experienced an airspeed range of zero to 130 KTS indicated airspeed, requiring an unusual attitude recovery.

Pilot advised Approach Control that we were descending to below 1,700 ft MSL as per original clearance. We were advised that we were east of the airport and to switch to Tower. On contacting Tower we were again advised we were well east of the airport and pilot-in-command requested ASR assistance while still heading northeast due to nasty weather pushing in from the west.

Tower switched us back to Approach Control, who again advised us that we were well east. By this time the pilot-in-command regained good visual ground reference at 400 ft MSL and elected to land at this location in a pasture.

It was impossible to communicate on the radio with Approach or Company Dispatchers to advise them of the safe landing.

While shutting down the helicopter, the pilot contacted Company Dispatch on his cell phone to report being safely on the ground and requested they advise the airport Tower and Approach Control. He then secured the helicopter and contacted the other pilot at his base to report safe.

NOTES:

GUSTY WINDS VS ROTOR
CALLBACK

MD500, Ferry, Phase: Parked, Name Withheld, Dec 2008

My Chief Pilot and I were on a ferry flight to the company office in ZZZ. This was day two of the trip as weather hampered our flight the day before. The conditions were clear but the winds were gusty.

We landed on our second leg at 1350 in ZZZ for fuel. After landing and shutting down, the airport representative discovered a padlock on the Jet A fuel controls preventing dispensing of the fuel. After waiting approximately one hour for the Airport Manager to return phone calls, a decision was made, after consulting our company mechanic and reviewing the procedures for 'emergency fuels' in the POH to accept a tank of 100LL and continue the trip.

We departed westerly and turned south to begin climbing to cruise altitude when at approximately 4000 ft we received an 'engine chips' light two miles from the airport. I immediately reduced power and turned back for the airport.

After a successful landing to a hover, my Chief Pilot instructed me to park the aircraft out of the way on a remote pad just off the ramp. This spot was chosen in the event the aircraft had to be left at the field, it would be out of the way and not disrupt airport Operations.

The engine was allowed its two minute cool down, per the POH, and shut down. With the battery turned off and rotors still spinning down, we both unbuckled, removed our flight helmets, exited the aircraft and walked to the rear of the fuselage to open the engine bay doors.

A quick visual inspection of the engine showed no indication for the cautionary warning and we were attempting to secure the doors in the open position when a loud bang was heard and the aircraft shook.

I immediately noted that the rotors were still turning, however, two of the main rotor blades were bent significantly. I then turned my attention in the direction of the tail where I noticed the tail boom on the left side had sustained damage — the gusting winds had pushed the spinning blades down and into the tail boom causing the damage.

The Director of Maintenance was notified and he later made arrangements to pick the aircraft up and trailer it back to the shop. Additional arrangements were made to have another company employee pick us up in a car, and the aircraft was pushed via ground handling wheels into a hangar where it could be secured.

NOTES:

CHAPTER 6

COMMUNICATE & AIR CREW

"The first real flying machine — that's what some people call a helicopter."

Ahnstrom, D.N.

STEEP DARK LEARNING CURVE
CASA

Unknown, Commercial, Phase: Cruise, Name withheld, Mar 2014

Your instructor may not always be right...

During my flying career I have had many memorable experiences, and learnt more valuable lessons than I can count. One experience that stands out from the rest was my 'working trip' to the United States. Three years into my new career as a helicopter pilot I was offered a position flying heavy multi-engine helicopters in the southeastern United States. Needless to say, I jumped at the opportunity – the chance of a lifetime! During the trip I gained experience working alongside American pilots, with a different airspace, culture and set of rules. While doing my aircraft conversion I learnt one valuable lesson that only really sank in properly years after I had come back to Australia.

Because of the type of work the organisation did I was required to fly low level using night vision goggles. I had successfully completed my day and IF conversion and was on my second-last NVG flight before my final handling test. I was finally becoming comfortable with the new aircraft and procedures. My instructor and I had twenty minutes of training left before we returned to the airfield and called it a night. All was going well until the master caution lit up – indicating the number two engine chip detector. (The chip detector is a small magnetic sensor located in the oil lines that detects small metal shavings in the oil. Worst-case scenario is a catastrophic malfunction somewhere inside the engine.) As the light was only flickering my instructor elected to continue the training. This was contrary to what I had been used to with my operator in Australia, but I told myself 'you're not at home; he's an instructor; he knows what he is doing; and he certainly knows the rules here better than you do!'

A few minutes later I noticed that the oil pressure on the number two engine was slowly falling out of limits. I announced it and he told me not to worry. The oil pressure finally fell outside of limits and we pulled the checklist and shut down the engine. The last checklist action was to land as soon as practicable. If I had been flying for my operator in Australia, this would have meant landing at the closest suitable airfield.

We were already established in the circuit of a satellite training airfield twenty minutes from home, and all my training and experience told me to land at this satellite field. However, my instructor had other ideas. He was going to fly home single engine, 'just like a big Jet Ranger'.

I knew that the trip home was over heavily vegetated undulating terrain, with no suitable landing areas in the event of a further malfunction. I was uncomfortable with this idea, as I believed it had unnecessary risks attached, especially when we were already right over a suitable airfield.

Rather than be assertive and speak up, I hinted at my discomfort with the plan by light-heartedly saying, 'my boss wouldn't be happy if I pressed on OEI (one engine inoperative), especially on goggles at 1000 feet. What if the other one lets go?' He laughingly told me 'ya'll Aussies are so safe; if one fails I'll just restart two!' He then assured me that he would get it going in 1000 feet. The voice in my head was back, telling me, 'he's an instructor; he knows the rules; it's his train set'. I kept quiet and cracked on. It was a long, nerve-wracking but eventually uneventful flight home. We declared an emergency at the boundary, and were met by a fleet of fire trucks on landing.

Lessons Learned:

Years later, upon reflection and with the benefit of experience and hindsight, I look back and wonder 'what if?' I checked my flight manual the day after the incident and was shocked to read that, 'in the event of a dual engine failure allow 6000 feet for a single engine relight'! At 1000 feet on NVG, over wooded, mountainous terrain, we wouldn't have stood a chance.

Even as a student, new hire, operating under unfamiliar rules, with an experienced instructor, you are a member of the crew. You are responsible for the safety of the flight as much as the captain. If nothing else, it's your backside strapped into the machine. It's easy to study human factors, to do CRM courses every three years; but is quite another thing to have the courage to put all the theory into action.

In hindsight, I'm lucky that the worst outcome for me is that I look back at that night with embarrassment every time I do a CRM refresher. That (now more mature) voice in the back of my head reminds me that by not speaking up I removed a countermeasure specifically designed to prevent an accident.

NOTES:

WHO'S GOT THE AIRCRAFT?
CALLBACK

Unknown type, Commercial, Phase: Cruise, Name withheld, Jan 2012

The Pilot flying was in the right seat. I was the Pilot not flying in the left seat with my head down energizing and programming equipment.

We were cleared to cross below the approach path from west to east and, "Report traffic on final in sight."

The Pilot Flying asked me, "You got the aircraft?"

I said, "Uhh, yeah" and took over the flight controls.

A Cessna 412 broke out of the overcast and apparently did not see [our] helicopter. We took no evasive action.

The Cessna cleared the top of the helicopter by just a few feet.

The right-seat pilot remarked, "Wow, that was close."

I responded, "What was close?"

I never saw the Cessna.

I thought the right-seat pilot wanted me to take control for some reason. He thought that I was confirming that I saw the conflicting aircraft and would take evasive action.

Lessons Learned:

Saying, "You got the aircraft?" only confused the issue. "Do you have the traffic?" would have been a better way to say it.

NOTES:

INTO THE VOID
CASA

R44, Commercial, Phase: Cruise, Name withheld, Jun 2006

Many commercial pilots can point to a single experience that changed their whole approach to flying. For me, that experience came when I had 250 hours in my logbook and I am grateful the lessons I learnt did not come at the cost of three lives.

I was scheduled to fly from Hervey Bay to Rockhampton in a four-seat Robinson R44. The Rockhampton agricultural show was underway and the aim of the trip was to establish a temporary joy-flight service for the duration of the show, a practice that had proven profitable for the company in the past. My boss was at the show and he had mentioned that he was keen for me to get there and start flying as soon as possible.

Two colleagues were joining me for the flight. While they were both ground crew, one of them also held a commercial helicopter pilot's licence. In fact, he had considerably more flight time than me though he hadn't flown for ages due to a long illness. Now recovered, it was agreed that this flight could be used to give him much needed co-pilot time. (I didn't realise it at the time, but he had flown only three hours in the past two years.)

By the time we departed, low cloud and fog extended along the coast and inland for 20nm. While the local weather looked poor, I was somewhat comforted by reports that the weather at our destination was near perfect; light winds and CAVOK. I took off and climbed to 500 ft, just below the cloud base. Visibility was about 5,000m. Once the aircraft was established on track – 5nm inland and parallel the coast – I handed over to the second pilot.

Visibility, which had been around 5,000 m at takeoff, now seemed to be deteriorating, making VFR navigation increasingly difficult. I asked the other pilot to track inland to avoid the worst of the weather. While we had plenty of fuel, the diversion would add a further half-an-hour to our flight time and possibly raise the ire of the owner, but I figured it was the safest course.

The second pilot disagreed with my decision, arguing that it would be better to track to the coast. Being less experienced and not wanting to tread on anyone's toes, I allowed his decision to stand. I didn't realise it at the time, but from that point on my role as pilot-in-command was compromised. We soon arrived over the coast, 300 ft above a straight piece of shoreline. According to my calculations, we should have been near a series of inlets. There were none, and I foolishly reasoned that the WAC chart had to be incorrect. We were flying into wind and were now 200ft above the ground, Visibility was appalling. Worse, with vegetation butting up against the shoreline, there was nowhere to land. I was still considering our options when I suddenly spotted a sand bar about 10m from the beach, 50m long and 20m wide.

I told the pilot to land, but he refused. He argued that the machine could be damaged if we were on the ground when the tide came in. I repeated the instruction to land, with an assurance that I would take the blame if the skids got wet. Again he refused. We continued along the coast, getting progressively lower and slower to stay clear of the falling cloud base.

As the conditions worsened my colleague acknowledged the seriousness of our predicament and agreed to turn back to the sand bar. My eyes were bulging trying to maintain sight of the ground. Then it was gone. For five or six seconds, we peered blindly into the muck, desperately hoping for a glimpse of the mangrove swamps just below us.

Then I saw the ocean: we were pitched forward 20 degrees, banking right at 45, and descending rapidly.

"We're too low!" I shouted.

He flared sharply and put the collective in his armpit.

"Low RPM!" was the next shout as the rotor speed decayed rapidly with the flare then dropped to the low 90s with the load placed on it.

He recovered RPM and, obviously shaken, started to land the helicopter in a mangrove swamp. I opened my door and was watching salt water rise up the skids getting closer to the air filter.

"What are you doing? We can't land here."

He pulled the machine out of the swamp then followed the coast till we found the nearby sandbar we'd passed earlier.

After landing, the second pilot relinquished any further control of the aircraft for the remainder of the trip. Then a break in the clouds exposed some blue sky revealing our location in a natural bay, and I was able to ascertain our exact position. Wanting to take advantage of the break we climbed aboard again and set off across the bay.

We called the chief pilot and explained the situation to him. He called the boss and told him there would be no more flying today. We completed the journey and two successful days of joy flight operations without incident.

Lessons Learned:

CASA Comment: I wonder how many pilots would admit to being in a situation like this one? At least one of the two pilots involved in this incident learnt an important lesson, and by sharing the story he has given other pilots the opportunity to benefit from his experience. Not all VFR-into-IMC situations end so well.

A useful tool for all pilots to remember is the acronym PPPPP – the five Ps – which stand for, "Prior planning prevents poor performance".

Although this was supposed to be a VFR flight, it's unclear whether the pilot prepared a flight plan, obtained a weather forecast or briefed his passengers before departure. At the very least, an area forecast would have given the pilot an indication of the likely weather conditions enroute. Armed with that information, he could have planned accordingly or, if required, postponed the flight. And by including the co-pilot and even the chief pilot in the flight planning process, any possible differences of opinion could have been resolved on the ground well before takeoff.

There can only be one captain in an aircraft. Without an appropriate understanding of the roles of the pilots on board, an adverse authority gradient can lead, as it did in this case, to confusion and misunderstanding. This is something the airlines have long known, and under the banner of crew resource management (CRM) airlines have honed their systems and procedures to ensure that crews clearly understand their roles and responsibilities and are adept at working as a team to achieve the best possible safety outcomes.

It would seem this flight departed with both pilots unsure of their roles. A detailed briefing before flight – covering at the very least who is in command, how tasks will be shared, and who will do what in the event of an emergency – would have averted many of the problems that these pilots encountered. Even if the second pilot's skills had been put to use helping with navigation and decision-making the outcome would have been markedly better.

Then there is the matter of "presson-itis". The decision by one of the pilots to press on into deteriorating weather – presumably to save time and to avoid delays – jeopardised the safety of the flight. It also resulted in them arriving at their destination later than they would have if the pilot had delayed departure, diverted, or put down at an early opportunity in accordance with the requests of his colleague.

This story is worth remembering if you are ever tempted to break the rules. Fortunately, this poor performance did not end in tragedy.

Prior planning should prevent a repeat performance.

Mal Walker
CASA Flying Operations Inspector

NOTES:

A NON-IFR CERTIFIED AIRCRAFT
CALLBACK

R44, Training, Phase: Initial Climb, Name Withheld, Feb 2014

I am currently training for my instrument rating. I filed an IFR flight plan requesting 2,000 ft enroute and in the remarks I listed no STAR, IFR training flight under VFR conditions, and that I would like vectors ZZZ. I was cleared to depart on the [SID] and then passed to Departure...where I was instructed to climb runway heading... and maintain 3,000 ft which I complied. Then I was instructed to turn right (do not remember the exact course but it was westbound), climb and maintain 4,000 FT. At this point I am flying over a cloud but my CFII still has visual reference to the ground ahead; I am in simulated IMC wearing a hood. We are then instructed to descend to 2,000 ft for separation, to which I respond, "Unable due to weather."

The Controller seemed puzzled that I wasn't able to descend through the cloud; I mentioned to the Controller that in the remarks of my flight plan I indicated the flight must be completed under VMC. The R-44 is not certified for IFR flight nor is it equipped for flight into known icing and I had been advised during my weather briefing of possible icing conditions above 1,000 MSL.

The Controller adamantly advised I must either comply with his instruction or declare an emergency. I responded saying that I did not need to declare an emergency but that I could not descend through the cloud. He replied with the same advisement.

By this time my CFII had informed me that we were clear of the cloud and able to descend; I immediately began descending and informed the Controller of the descent. I assume he didn't hear my transmission because he repeated his previous advice for a third time. I again informed him I was descending to 2,000 ft in compliance with his instruction. We were then passed off to another Controller and not long after again we were given a vector that would fly us into IMC and I advised that I was unable and needed to climb to avoid a cloud. I began climbing to avoid the cloud. The Controller then advised me that I may not deviate from ATC instruction and gave me a phone number to call for a possible pilot deviation once we had landed.

My CFII did tell me that when she called the number the person she spoke with informed her that we were not allowed to file and fly an IFR flight plan in an aircraft not capable of IFR even for training purposes which was unknown to virtually everyone we have come into contact with regarding the occurrence.

Lessons Learned:

Looking back, and after several conversations with other instructors, it seems the best course of action would have been to cancel our IFR clearance as soon as we were instructed to do something that would violate our operating limitations since we were in VMC at that time.

Perhaps due to inexperience on my part, that hadn't even occurred to me. In my mind it seemed either decision would have been a violation of some sort.

Deviation from ATC is a violation, but flying into IMC and possible known icing in an R-44 helicopter could have resulted in an accident with loss of life, both for me and my instructor but also civilian casualties below in the populated suburbs.

When weighing those options in the cockpit it seemed to me that avoiding the cloud after informing ATC of my situation was the safer course of action.

<u>NOTES:</u>

MAINTENANCE REPAIR FACILITY
CALLBACK

S-76, Passenger, Phase: Maintenance, Name Withheld, Aug 2013

Below is a report sent to Company X Corporate Flight Department in response to two maintenance issues and current Operating Policy. Followed up with my resignation. I would first like to discuss the events that surround the current issue related to Aircraft X, a Sikorsky S-76 helicopter, followed by my observations related to the damper replacement one week earlier on Aircraft Y; and observations related to tool control, hardware control and safety policy. The intent of this report is not meant to be malicious but viewed as an opportunity to improve your safety process.

August 2013, I reinstalled the Number-Two Forward Firewall on Aircraft X; the first panel that I installed had several camlocks that required replacement. The camlocks were in a parts bag and there were no retainers for the camlocks, they were not mentioned in the turnover notes nor was there any mention of the disposal of the retainers. I inspected the area for the missing retainers and requested Quality Assurance (Q/A) also inspect the area prior to install of the panel; no hardware was found. The missing retainers were mentioned to Q/A, I don't remember the exact response but it was not one of concern. For the installation of the second panel I started by opening the hardware bag and counting the hardware and then counting the number of mounting points on the part. I found that I had three extra bolts and was missing one spacer and four washers. This was brought to the attention of QA and the response was " I will ask Inspector X about it when he comes in". I received a QA approval to install the panel, the panel was installed minus one bolt at the lower outboard corner, there were several attempts made by Inspector Y and myself to start the bolt and we concluded the nutplate was damaged and would need to be replaced at a later date. I entered a discrepancy in the Aircraft Maintenance book and signed the panel install. Not being sure of what to do with the extra hardware I placed it back in a bag and placed it in my desk knowing that I would need this hardware at a later time. When I left that afternoon, Inspectors X and Y were preparing to start the scheduled aircraft inspections. One hour later I received a call asking me for the location of the hardware; I told them exactly where it was. The next day I checked the aircraft, the missing bolt was installed and I could not find a record of the discrepancy that I had entered. I asked Mr. X what they did to fix it and the response was "we bent the tab". This was followed up by a morning teleconference with Mr.X and the Director of Operations. The purpose of the call was to lecture me about the hardware that was placed in my desk.

The previous week a Main Rotor Damper was replaced on another S76 helicopter; the tools that were used to perform the task were out of a personal tool box and there was no tech data readily available, this is a major violation of tool control. That week in the staff meeting there was mention of knowledge of the uncontrolled tool box and that it needed to be removed. The tool control at Company X requires improvement. The Tool Box should be kept in a secured area when not in use and a record of who used the tool box and on what aircraft it should be maintained. Tool Box should be inventoried when signed-out and in and should have an inventory of what is in each drawer that alerts the individual as to exactly what they are looking for if a tool is missing. All the junk in the bottom drawer of the tool box should be removed because it is uncontrolled and could be considered FOD.

During the aircraft post flight inspection, the fuel samples are taken; the proper time to take fuel samples is during pre-flight or when the aircraft has remained static for a minimum of one hour. The fuel samples for the Fuel Farm should be done at the beginning of the day prior to any aircraft being fuelled. Hardware needs to be controlled, and the Hardware Crib access should be restricted. When new hardware is required the mechanic should provide a part number for the hardware and when new hardware is issued, the old hardware should be properly disposed of so it does not become FOD. The Consumable Cabinets require an inventory sheet and the expiration dates need to be reviewed. There needs to be a list of what material is flammable and stored in the proper cabinet. One afternoon when I came in, the facilities person came into the office and pointed out that there was still flammable materials being stored in the wrong cabinet. The response from the Maintenance crew was "OK, thanks, we will take care of it" and they went back to what they were doing. They should have made a note or created an action item list. The caulking and the HYSOL [adhesive] in the refrigerator needs to be removed because it is a health issue and does not require refrigeration. When there is a lost tool or missing hardware there needs to be a process in place for grounding the aircraft until the missing item is found or a satisfactory inspection has been completed. If your safety policy is going to work you need to make it clear that when there are missing items reported, there will be no disciplinary action.

Lessons Learned:

CALLBACK Comment: Reporter stated the Main Rotor Damper attach bolt uses a castellated bolt and nut assembly with a cotter pin. The S-76 had diverted to another airport due to vibration issues and the fact that, there were company Maintenance personnel available.

Reporter states he checked off the 'Briefing' item on the ASRS form to reference the lack of Shift Turnover briefings describing the maintenance work that was 'in progress'. The lack of adequate record keeping and accountability of tools used and who removed a tool from their Tool Shadow boxes, continues to be a problem. He has since left the company to work for an MRO Contract Repair Station that seems to be more focused on doing maintenance work correctly.

NOTES:

PUSH DOWN THE COLLECTIVE
CALLBACK

Eurocopter AS 350, Training, Phase: Landing, June 2009

I was conducting the initial night vision goggle (NVG) training for a company pilot. Flight manoeuvres focused on normal takeoffs and landings, hover work and hover taxi with emphasis placed on the benefits and limitations associated with NVGs. He stated his biggest challenge was gauging his skid height above the ground in a hover.

I allowed him several takeoffs and landings to and from a hover to get accustomed. He shared with me that he wore corrective lenses for nearsightedness, information that I failed to place enough importance to as I think it contributed to our hard landing.

After two hours of training I had the training pilot place the helicopter in a stable 3' hover and, as per our brief, I announced and initiated a simulated engine failure by retarding the fuel flow control lever (throttle) toward the idle/cutoff position.

Contrary to procedures the training pilot immediately increased collective pitch resulting in a climb and decayed rotor rpm. I countered by pushing down on the collective and telling him to lower the collective in an effort to save rpm.

As we settled I commanded, "Pull collective" and pulled to cushion our landing. However, he continued to push down on the collective. We impacted in a level attitude with force sufficient to activate the ELT. Neither pilot sustained injury. I directed the training pilot to shut down the helicopter. We then exited the aircraft to inspect for damage. I found obvious deformities to the tail boom in three locations. Extent of the damage is yet to be determined.

Factors that contributed to this incident:

The training pilot's corrective lenses are fairly small and wire framed. Given that he is nearsighted, it is probable that he does not have optimum vision to either side of and below the NVG tubes. Even with my confirmation that we were at the correct skid height to initiate a hovering auto rotation, his response was to immediately increase collective pitch. I have no explanation as to why he countered me on the controls.

He later stated that he did not want to raise the collective because the low RPM warning horn had been on too long. Insufficient time to affect a positive transfer of controls complicated the situation.

While we would still have landed firmly, additional collective input could possibly have cushioned the landing enough to avoid damage to the helicopter. The manoeuvre is brief and even under ideal conditions a positive transfer would have to be very flawlessly executed let alone when the manoeuvre is improperly initiated. Disagreement in control input is probably the greatest contributing factor to this incident.

Lessons Learned:

Greater emphasis on the effects of corrective lenses to the wearer. I will be sure I know and appreciate the significance of their impact. Add discussion to the preflight brief and the manoeuvre description addressing the potential for transfer of controls if things go awry.

<u>NOTES:</u>

CHAPTER 7

NEAR MISS & SEPARATION

"One of the most important functions of this aircraft, the flying ambulance. Medevac helicopters (medical evacuation) help save thousands of lives per year."

Modern Marvels
History Channel

DAUPHIN VS SKYHAWK
CALLBACK

SA 360 Dauphin vs Skyhawk 172, Military & Private, Phase: Final Approach, Dec 2013

MH-65C operating in VMC conditions on an IFR flight plan accepted a best speed approach at 120 KIAS from ATC due to converging fixed wing traffic in trail on VOR/DME Runway 15 approach.

At approximately 8 NM north of the field, ATC cleared MH-65C to switch to CTAF and canceled radar services. In error, pilot not flying tuned incorrect CTAF in COMM 1 and announced position and intentions to perform a go-around at the approach end of Runway 15 followed by a departure to the south.

At an estimated 2 NM north of the field at 880 ft MSL, the Flight Mechanic noticed and announced a shadow of an aircraft projected on the ground in the vicinity of the departure end of Runway 33.

As the pilot not flying and pilot flying began searching for traffic, the Flight Mechanic announced "break left."

The pilot flying immediately performed a descending left turn at an estimated 30 degrees angle of bank to avoid the oncoming traffic. The single engine, general aviation aircraft passed left to right at an estimated distance of 400 ft.

The pilot not flying immediately realized the wrong frequency had been entered into COMM 1, tuned the correct CTAF, and announced intentions. The MH-65 continued flight on its original IFR flight plan without further incident.

The pilot not flying reported developing a habit pattern of not utilizing the Preset frequency when in the left seat since that practice requires reaching across the cockpit and putting the flight helmet in close proximity to the FADEC control switches. Prior to tuning the CTAF, the pilot not flying verified that the pilot flying had tuned the proper frequency in the Preset position.

When instructed by ATC, and in an attempt to quickly change to the CTAF frequency, the pilot not flying reverted back to an old habit pattern and selected the COMM1 right line select key (instead of the Preset button) which made the CDU "buried" frequency the new COMM1 active frequency.

The pilot not flying admittedly felt rushed due to accepting the best speed approach at 120 KIAS and failed to verify the correct frequency had been entered into COMM 1 after depressing the right line select on the CDU. No aural TCAS alert was heard by the crew prior to the mishap.

The TCAS tested properly on deck and exhibited normal operation throughout the flight.

It is suspected that the general aviation aircraft was either not equipped or not operating with a transponder. The crew was unable to establish good communications with the general aviation aircraft due to a heavy workload immediately following the mishap.

After the flight, it was noted that the field is utilized heavily by a local flight school. It is unusual for [our] unit aircraft to operate in and out of this area.

This was the first time the crew had been to [this field].

Lessons Learned:

There are two valuable lessons we can learn from this mishap:

1. The first is the utmost importance of the Flight Mechanic maintaining a good outside visual scan during IFR training flights. Had the Flight Mechanic not seen the oncoming aircraft and had the Flight Mechanic not been accurate, bold, and concise while commanding the pilot flying to "break left", we could have had lost an aircraft and crew.
2. The second lesson is that we had a crew experience a loss of situational awareness because they were rushing an instrument approach.

Often times, you'll find yourself slowing down the pattern as a helicopter while operating in the IFR environment. This is a situation that can be stressful if you unnecessarily impose pressure on yourself to comply with optional ATC requests.

Don't accept more risk than you should for the convenience of ATC or the aircraft behind you. If you are uncomfortable with the request, simply state that you are "unable". ATC is not always fully aware of your capabilities and it's your responsibility as an aviator to prevent ATC from getting you in a bad situation behind the aircraft.

<u>NOTES:</u>

POLICE VS EMS
CALLBACK

MD-500MG vs AS 350, Utility & Ambulance, Phase: Cruise, Names Withheld, Mar 2013

I was PIC in a police helicopter when I took evasive action to avoid a near collision with an EMS helicopter while in the Class Delta airspace.

I requested permission to enter the Class Delta airspace from Tower and was granted permission. I was at 1,900 FT MSL and on an approximate heading of 055 degrees and 80 KIAS. Tower advised me of an EMS helicopter in my vicinity to the east and westbound through the airspace. I advised the Tower I would be looking for the traffic. I heard Tower advise the other aircraft that I would be transitioning in the airspace also northeast bound. I hear the EMS pilot acknowledge the Tower and state he would be looking for me as well.

We received no further warnings or updates from the Tower Controller.

I began scanning the horizon for the westbound helicopter and periodically checking the Garmin 530 display for any traffic targets in that area. The Garmin showed no targets each time I checked it.

After approximately 45 seconds to one minute of scanning without seeing the converging aircraft, my Observer pointed out the aircraft at my 11 o'clock position. The yellow AS350 helicopter was westbound and approximately 400 ft horizontal distance and nearly the same altitude if only slightly higher.

I took evasive action by turning right approximately 10 degrees quickly and kept my altitude of 1,900 ft MSL. It did not appear that the yellow EMS helicopter deviated in any way on its course. The EMS pilot did not report having me in sight to the Tower and subsequently left the Class Delta airspace to the west. I continued my northeast bound course without incident.

[In summary] two helicopters on opposite course at high rate of closure. Neither pilot seeing the other until the last moment. Traffic avoidance equipment in one helicopter [was] not displaying targets. Pilot of aircraft X assumed other aircraft was further away than it actually was based on ATC's description of where aircraft was at time of advisory warning.

No visual identification made by either pilot. No follow-up from ATC as aircraft converged at nearly same altitude. Traffic advisory equipment/Garmin 530 [was] not working at time of incident.

Lessons Learned:

Ask for further update from ATC or advise "No joy" on other aircraft as time passes without seeing the other aircraft. Ask crew member to assist in looking for other aircraft.

<u>NOTES:</u>

BELL VS TWIN
CALLBACK

Bell 222, Training, Phase: Final Approach, Name Withheld, July 2012

During a helicopter training exercise (2 pilots) on a 45 degree to runway short final Runway 17, landing with a simulated hydraulic system failure, we made a last minute check to make sure no one was landing on the runway. Upon looking to the north saw a light twin on very short final for Runway 17. Both aircraft were closing in on the approach end of the runway at the same height and distance for landing. The twin was to our right and slightly behind our field of vision about 200 ft and closing.

Once the other aircraft was spotted we terminated the approach to a high hover and the Twin continued with its landing. There was no further incident.

Lessons Learned:

Communications would be the main contributing factor.

The airport was very quiet and we had been monitoring

Approach Control and the local CTAF frequency as well as making our own traffic calls. We never heard any radio calls from the twin until just before he departed the airfield.

I also believe TCAS in either aircraft would have been helpful in preventing this. The fact that Texas Gulf Coast Regional (LBX) and Houston Southwest (AXH) share the same frequency and are very close to each other and the high amount of radio calls at Houston Southwest may have contributed to the confusion.

NOTES:

BIRD VS WINDSCREEN

CALLBACK

Bell Helicopter, Ambulance, Phase : Cruise, Name Withheld, Sept 2010

On the second (patient) leg of the flight, while the aircraft was in cruise flight, a bird flew into the pilot's windshield, breaking the windshield and leaving a 7 inch hole in the windscreen.

The Aircraft was enroute to a hospital. We were 35 minutes into the flight with 45 minutes remaining. The aircraft was at 8,000 MSL and cruising at 110 KIAS and 125 KTS ground speed. The pilot was on NVGs (night vision goggles) and the medical crew had their NVGs removed for patient care.

Numerous birds had been observed through the flight.

After the bird struck the windscreen airspeed was slowed to 80 KIAS.

After determining that everyone was unhurt the damage on the windscreen was assessed. The hole was beside the center divider and just below the dash visor. It was determined that the windshield was structurally sound and the closest safe landing area was back behind us 25 minutes away at 80 KIAS. The pilot informed dispatch that we had been struck by a bird, which went through the windscreen. We also informed them that no injuries occurred and we would be diverting. Dispatch notified the hospital that there would be a 1 hour delay on patient arrival.

The medical crew contacted our fixed wing aircraft, which was returning to its base at our diversion airport. It was determined that it would land at the airport 10 minutes after we would land. We flew to the diversion airport and made a normal approach and hover taxied to our parking area and shut down the aircraft.

Aviation and medical management was notified of the event and our safe arrival. They were also notified of arrangements made to transport the patient. The aircraft was then inspected for any other damage with none being found.

A write up was placed in the aircraft log book and the red sock was placed over the cyclic. All medical equipment and publications were removed from aircraft in anticipation of our primary aircraft going back into service the following day. Once we got back to the hospital, which is were the helicopter is based, the equipment was placed in the hangar. The pilot then debriefed the crew and called and debriefed the Dispatcher.

Moon had set at XA:30. Sky was clear visibility was good, with moonlight being very low.

Lessons Learned:

Deviate route or altitude if possible to avoid birds.

NOTES:

R44 VS UAV
CALLBACK

R44, Commercial, Phase: Landing, Name Withheld, Sept 2018

I was dropping off a groom and his best man for a wedding at a venue with a tight landing area. It was a short flight, and after communicating with Tower I began my high recon of the landing area. I had done both a satellite imagery review and a site visit prior to the landing, but there was a new obstacle I had not expected to encounter. One of the groom's friends had a drone that he was hovering in the parking lot. Not 100% sure of the type but it was a small four rotor system similar to DJI Phantom with a camera underneath.

I had not thought to communicate to the groom ahead of time of the need to keep any small UAS on the ground during the landing. I did say to the groom during the flight, "He needs to keep the drone on the ground."

As I shifted my focus back down to the ground I saw the UAS was in his hand and it looked like he was walking it back to his trunk. At the same time the groom was on the phone and I heard him say, "The pilot said to keep it on the ground."

I decided at that time to continue the approach, and conducted a steep approach into the landing area to remain clear of the trees and obstacles in the area. After landing, I rolled down the throttle to bring rotor RPM to idle and had the groom and his best man exit out of the helicopter walking forward of the helicopter. When I shifted my attention back forward I saw the drone back in the air, about 10 feet in front of me at or slightly above my rotor system. I leaned my head out of the aircraft and made eye contact with the operator while pointing at him, then the drone. I made a hand signal to back away from the aircraft, and the drone moved away from the helicopter and back down to a one foot hover before setting back down on the ground. At that point I contacted tower for takeoff clearance, brought my RPM back up to flight and exited the landing area using a max performance takeoff.

Lessons Learned:

It is possible communication with the groom prior to the event to keep any aerial videographers on the ground during the landing and takeoff would have prevented the occurrence, but it is possible the drone operator never communicated his intent to film the landing to the groom.

What would have been far more effective would have been to have ground personnel there for the landing to directly communicate with the operator and stress the importance of keeping the drone on the ground to prevent either a mid-air collision or the drone being thrown by the rotor wash into people or objects.

However, we had limited staffing due to the holiday weekend and all available company personnel were tasked. I have no way of determining if the operator was licensed, I consider it a high probability the individual was a friend who flew for hobby. A factor in my assessment of this probability is the hope that a licensed UAS operator would know better than flying a UAS two miles from a class B Airport off the departure end of the runway.

NOTES:

INCORRECT FREQUENCY
CALLBACK

R44, Passenger, Phase: Cruise, Name Withheld, Jan 2018

While conducting tour operations I was headed south west bound along the coast line south of the city. I was flying a Robinson R44 with two passengers onboard. I was receiving radar services from Approach via a discreet transponder code that is outlined in our Letter of Agreement.

The turnaround point on the particular flight I was flying was the lighthouse. Typically, if I have wind coming from the south west, I will stay 200-300 ft AGL as I transition down the coast line south west bound to stay well below any traffic arriving to and departing [at the nearby airport]. If the wind is not a headwind I will raise my altitude accordingly.

On this particular flight, as was true most of the afternoon, I had approximately 12 knots of head wind and was cruising around 300 ft AGL. I maintained my altitude and slowed from cruise speed to my "Orbit speed" of between 60-70 KIAS as I approached the lighthouse. About midway through my left hand orbit of the light house, a Single Engine High Winged Cessna appeared from behind the lighthouse south bound in a left bank. It was level with my compass on collision course. I immediately lowered collective into the full down position, entered autorotation and lowered the nose to gain airspeed and increase my rate of descent. I recovered from autorotation approximately 100 ft AGL.

The airplane crossed what appeared to be directly above us and would have hit us had I not performed the evasive action. After the airplane flew over us we both continued our orbit. I had slowed down to approximately 35-40 knots and was using my anti torque pedals to fly out of trim to keep the airplane off of my nose and in my direct field of vision. The airplane didn't appear to have changed altitude and I passed behind him as he departed to the north east. I do not believe they saw me. I was unable to obtain his N number. About the time I finished my orbit and passed behind the airplane, I could hear the controller working Approach call my N number. I responded and told the controller that, "I'd like to report an NMAC."

He Responded, "Yeah, I saw that. Are you ok?"

Without going into the whole dialogue, I reported the Near Mid-Air Collision with the Cessna. I reported it as being either a 172 or 182 because I wasn't sure which type. It was white with a green stripe running down the side which had brown or tan accents on the stripe. The N number was in the stripe on the tail cone. I never established radio communication with that airplane.

Lessons Learned:

[Local] Approach is very good about giving me traffic advisories and does so on daily basis. It is possible that I did not hear the traffic Advisory or alert since I was monitoring Approach, Unicom at [another nearby airport], and Helicopter air to air frequency as per AIM 4-1-11. I was also conducting a tour and conversing with one of the passengers right before the event.

I think that the traffic may have been flying lower than what ATC's Radar would pick up and therefore didn't show up on his display to warn me until the airplane climbed just before reaching the lighthouse to make an orbit. I say this, because the airplane descended after our encounter, which Approach confirmed.

There is a BIG problem with low flying airplanes in the area where we conduct tours. Very few of them are on the radio from my experience in trying to reach them. I think it's partially because of the Class C airspace shelf beginning at 1,200 ft msl. For some strange reason, people are afraid to talk to ATC, or they just don't want to so they stay below it. People also want to fly over [local landmark] and over the lighthouses and over [the city] to see the sights. Almost all do so at an altitude below prescribed minimum safe altitudes prescribed in 91.119.

The airplane I encountered at the lighthouse was definitely within 500 ft of a structure (the lighthouse). I think another problem is that a lot of people come into the harbor from [another] airport. I would too. The fuel is cheaper there and it's an easier airport to navigate. Therefore, people are operating closer to an airport with a different frequency than the frequency to which they are tuned.

In my particular aircraft, I have the capabilities of monitoring 3 frequencies. I choose to monitor helicopter air to air, UNICOM, and Approach for Radar Services. The problem is that, that is both not enough frequencies and too many all at the same time. There are times when there is someone talking on all 3 frequencies at the same time and you have to decide which two to shut off. The company that I work for has pushed that we publish the air to air frequency on the VFR Sectional. Personally, I would like that whole area made an "Alert Area" with its own frequency printed on the chart. I think it would relieve radio congestion and give pilots guidance as to what frequency they should self-announce on.

I also believe that ADSB cannot come soon enough and needs to be mandatory for all aircraft with an onboard electrical system for operation in ALL airspace. I was unable to read this airplanes N number and was not able to get a hold of them on the radio. Had the airplane been equipped with ADSB out capabilities, Approach would have seen who it was, and I could have too if I had ADSB in capabilities. Because of this, I think that accountability will resolve many issues because people know they are being watched. And if they don't know that they're doing something wrong, they will be able to be educated, because they can be tracked down.

NOTES:

BELL VS BOEING
CALLBACK

Bell & B737, Ambulance & Passenger, Phase: Cruise, Name Withheld, June 2000

Aircraft X (Bell) had responded to a single vehicle rollover accident on the interstate. We departed the scene with 1 patient enroute to hospital. Approximately 19 nautical miles south east of ZZZ, I established initial radio contact with ZZZ approach, acknowledging ATIS and requesting lifeguard priority for hospital. Approach immediately assigned a mode c squawk. Approximately 15 nautical miles south east of ZZZ, the approach controller advised that we were radar contact and cleared us direct to hospital. Our altitude at this time was 1500 ft MSL based on the altimeter setting provided by ATIS, with a 140 KIAS ground speed indicated by GPS. We were heading approximately 310 degrees.

Approximately 10 nautical miles south east of ZZZ, the approach controller queried if we intended to 'cross over the ZZZ airport.' I advised the approach controller that we could accept a midfield crossing if necessary. The controller advised me to 'expect midfield crossing due to high traffic volume.' I then acknowledged this.

Approximately 2.5 nautical miles south east of ZZZ, the approach controller cleared us to 'cross midfield direct to hospital.' The controller advised that he had an aircraft on approach for runway 17L and an aircraft on approach for runway 17R, as well as one aircraft 'about to take off on runway 17R.' as I neared runway 17L, I initiated a descent to 1300 ft MSL. At that time, I could see an aircraft that was on about a 4 miles final for runway 17L as well as an additional aircraft slowly vacating runway 17L. I could also see an aircraft Z jet which was taxiing into takeoff position on runway 17R, as well as one aircraft on about a 6 mile final for runway 17R. I acknowledged having visual contact with the stated traffic.

After crossing the east runway, I started to adjust my course slightly left in order to bisect the west runway since I judged this to be the safest point of crossing. Aircraft Z had taxied into position and held on the runway for such duration that I initially thought he may have been given 'position and hold' instructions by tower to facilitate our runway crossing. As I started this l course correction, the aircraft Z jet which had been holding in position started his takeoff roll. I rolled level immediately having some concern that the point of rotation of the aircraft Z jet might place him in a climb toward our projected point of runway crossing. I looked to the right and could see an aircraft on approach, which had advanced to an approximately 3 mile final. I looked again to the left and witnessed the aircraft Z jet rotate. Since aircraft Z had rotated well to our left (south) and in front of us, his climb out was no longer a factor.

As I looked forward, my crew member yelled, "Down, down, down!" I placed the aircraft in a dive at the same time observing a B737 to our immediate right heading directly toward us. I started a left turn away from the Boeing, but then rolled level as I judged the turn might place the rotor system of our helicopter closer in proximity to the advancing Boeing. The B737 at my initial sighting was approximately 300 ft away, at our altitude in a climbing right turn. The B737 passed directly overhead with what I estimate to be 150 ft of vertical separation. It appeared that the B737 had increased his

right turn rate between my initial observance and his passing overhead, leading me to question whether the B737 had initiated evasive
action.

I immediately called ZZZ approach, however, did not receive a
reply. I made a total of 3 calls to ZZZ approach before receiving an
answer which consisted of 'Aircraft x radar service terminated,
squawk VFR, have a good flight.' The controller who had answered
and issued this instruction was a different controller than the one
who had been handling us until that point.

I explained to the controller that I needed to "landline" him upon
arrival at the hospital. Initially, he did not understand my request, but
eventually understood my need to speak to him by telephone. The
controller provided a telephone number. We continued our flight to
hospital without incident. There was no injury to crew or pax as a
result of our evasive manoeuvre.

After landing, I attempted to call the telephone number given,
but could not get through due to a busy signal. I immediately called
the aircraft X chief pilot and reported the near miss. He immediately
contacted the shift supervisor at the ZZZ ATC facility to relate the
incident. I subsequently spoke to the tower supervisor after returning
to our base approximately one hour later. I informed him that I had to
take evasive action to avoid a possible collision and reviewed the
events as I knew them. The tower supervisor indicated that to his
knowledge the aircraft Y, B737, crew was unaware of the nautical
near miss as they had not reported sighting our helicopter at all, and
had not initiated any evasive action. The supervisor indicated that
both the tower and approach controllers involved were unaware that
a near miss had occurred, and that the TCASii in the aircraft Y B737
and in the ATC facility had not given warning of the conflict. The
supervisor had already reviewed the audio tapes of the incident and
indicated that in his judgement, the ATC facility was completely at
fault for the traffic conflict. He stated that I was in no way responsible
for the near miss, therefore, no enforcement action would be
pursued. He indicated that he was initiating a 'quality assurance

review' to review the incident, and as a result of this process, aircraft X company would be informed of their findings.

Lessons Learned:

After thorough discussion with the tower supervisor and post flight review with my crew, I would like to offer the following as possible contributing factors to the incident:

1. Operator error by ZZZ ATC for not completing a formal heading of aircraft X from the approach controller to the tower controller. If a hdof had been executed, rather than a point out between the 2 controllers, then com between conflicting aircraft would have been facilitated. The point out is a common practice by ZZZ ATC, the efficacy of which is questionable, due to conflicting traffic being on different frequencies.

2. Operator error by ZZZ ATC for inadequate spacing of arriving traffic.

3. Operator error by ZZZ ATC for providing 'lifeguard priority' in name only without any actual priority being given to the lifeguard aircraft.

4. Controller error by ZZZ tower for improper traffic call to the aircraft Y B737 after gar instruction had been given. Tower controller advised aircraft X of 'helicopter traffic crossing left to right, below you, not a factor,' when in fact both aircraft were at the same altitude and converging.

5. Operator error by ZZZ ATC for changing controllers and combining radar East and radar West position during the critical moments preceding aircraft confliction.

6. Controller error by ZZZ Approach for not advising aircraft X of the gar issued to the aircraft Y B737.

7. Controller error by ZZZ Approach for not advising aircraft X that there were actually two aircraft rather than

one aircraft on approach for the w runway 17R.

8. Controller error by ZZZ ATC for terminating radar service immediately after crossed the west runway. There was still the departing aircraft Z jet in a climbing Right turn and the aircraft Y B737 in a climbing Right turn creating the continued threat of confliction.

9. Aircraft design of Bell 206L3 which limits forward visibility at the aircraft's 10 o'clock and 2 o'clock position due to a 6-7 inch wide door frame. A target must become fairly large and therefore quite close before this no longer obscures the target. Remedy is for pilot to not only turn the head, but also rock forward and aft in the seat to see around this obstruction.

10. Equipment failure of traffic collision avoidance system to alert ATC and aircraft crew of aircraft proximity.

11. Pilot error by aircraft Y B737 crew for not observing the aircraft X, helicopter, particularly after being made aware of our presence by the tower controller.

12. Pilot error on my part for not observing the aircraft Y B737 before it became a hazard. Callback conversation with reporter revealed the following info: reporter stated the facility changed their procedure in handling helicopter traffic crossing the airport by requiring the approach controller to make a hdof to the tower. He said that recently he has noticed the facility going back to their old ways of doing things and approach control is again only making point outs to the lcl controller. Because he only told the tower supervisor he had to take evasive action to avoid the B737 and did not request to file a report, the incident was never classified as a near miss.

NOTES:

IROQUOIS 212 VS BEECHCRAFT
CALLBACK

Bell 212 & Beechcraft, Air Taxi, Phase: Takeoff, July 2000

I called Grant County (Moses Lake, WA) Ground Control on 121.9 requesting, "South departure to Redmond, or, at or below 6500 ft", the tower/ground controller, being the same voice, responded with, "Stand by."

He called back approximately two minutes later asking what I wanted. I again asked for an S departure from the customs ramp.

Controller responded, "Squawk xxxx, departure frequency 126.4, contact tower on uh...12...Stand by...Are you ready to go now?...Stand by...Tower 118.25."

I was unable to read back these instructions for I heard the same controller on 118.25 directing traffic on both runway 18 and runway 32R in between the 'standbys.' There appeared to be 3 or 4 heavy Boeing airliners working runway 32R, 2 x military C5 or 117's and other light training aircraft on runway 18. I was monitoring both ground and tower on both VHF radios. I hover taxied n away from the customs box and turned left, started my takeoff to the southwest.

I called. "Aircraft x on the go."

Tower: "Where are you?"

Pilot: "On the go from the South West ramp climbing through 300 ft southbound."

Controller: "Traffic is a beech craft touch-and-goes runway 18."

Pilot of x: "I have the traffic, thanks."

Did I miss something? I saw the Beechcraft trainer at my 5 o'clock position, same altitude, 400 ft AGL, Approximately 500 ft or less horizon. Had the tower controller not pointed out the traffic, I feel that this could have resulted in a midair collision. I levelled at 400 ft AGL, turned slightly l to avoid traffic. I believe that the Beechcraft was totally unaware of my position. He appeared to continued his climb. I continued at approximately 400 ft AGL for several mi, asked the controller the position of the Beechcraft, being unable to see him behind me after I turned l and levelled off.

I was told he was in the pattern of runway 18 for touch-and-goes, contact departure on 126.4. I was soon turned over to chinook Approach on 128.75.

Wow, why a close encounter with a Beechcraft?

Problems: ground controller was working both ground control, giving clearances, working the w tower frequency with 7 or more aircraft. It is my impression that the controller was not able to give me clear departure instructions. I was not able to read back these instructions when given. I did attempt to 'read back' on our third or fourth gar. And then I felt rushed by the comment 'Are you ready to go now?' on the ground frequency.

1) Were instructions clearly given by controller?

Pilot: No.

2) Was my aircraft given proper and equal attention as the heavy Boeing jets and other trainers doing touch-and-goes on runway 32R and runway 18?

Pilot: No.

3) Did the pilot of the helicopter feel pressured for an immediate takeoff from controller?

Pilot: Yes.

Being a dual rated pilot in both airplanes and helicopters, I have experienced that many controllers (including the one mentioned above) may view helicopters with discontent and in many instances will not allow 'us' helicopter pilots the same consideration as fixed wing aircraft. I felt that this attitude was prevalent on this date and led me to a state of frustration and distress. This was a major contributing factor to this possible midair collision.

Factors:

- Pilots feeling on not receiving equal attention from controller.
- Pilots frustration with controller leading to impatience and distraction from takeoff instructions.
- Possible controller being overtasked in running both ground control, issuing clearances, and tower control with many aircraft requesting attention at the same time.
- Controller not insisting on a 'read back' on takeoff instructions.
- Ambiguous question 'Are you ready now?' on ground control.
- Pilots feeling of pressure to take off now.
- Pilots not stepping back, taking a breath, and saying 'Wait a min, I will not go until I am good and ready for takeoff.'
- Not having a separate ground/clearance delivery controller (free from tower control responsibilities) that can give full clearance and/or ground control instructions with proper read backs from pilots.

Lessons Learned:

What worked: Pilot called: "Call sign on the go south bound." Controller: "Where are you call sign?" Pilot: "On the go South West ramp climbing through 300 ft." Controller: "Traffic is a Beechcraft touch-and-goes runway 18." (Had he not mentioned this I hate to think of the possibilities.)

Although I fault the controller on some issues, he was quick to take charge and point out traffic at just the right time.

The FARS allow helicopter pilots much latitude in approaching and departing airports. This works very well most of the time except when there are noise abatement issues and when the traffic pattern is congested. This (congestion) raises the stakes considerable. It is then up to the helicopter pilot to access the risks of 'avoid the flow of fixed wing' and mix when it is in the best interest of all involved.

Yes, that may mean creating more workload for the controller, but consider the risks.

NOTES:

CHAPTER 8

KNOWLEDGE, DECISION MAKING & RISK

"On June 21, 1972, in an Aerospatiale SA-315B Lama that was lightened as much as possible, I reached an altitude of 40,814 feet establishing a helicopter altitude which remains today. That flight also wound up being the longest autorotation in history because the turbine died as soon as I reduced power. With a -63°C temperature that day, the engine flamed out and could not be restarted."

Jean Boulet, Aerospatiale Test Pilot.

BAD VIBRATIONS
CASA

Hughes 500C, Commercial, Phase: Cruise, Name withheld, Dec 2006

Sometimes the signs of looming danger are subtle and "it just doesn't feel right" is the closest you can get to pin-pointing a problem.

That was the case on Valentine's Day 2000. I was working as a helicopter pilot on a Korean fishing boat in the south-west Pacific. The job of helicopter pilots on these vessels is to fly up to 60nm away from the boat to look for signs of feeding tuna. If you spot a school, you radio its location and start herding the school while the net is being set. Other tasks include collecting beer, mail, videos, spare parts and transferring personnel on and off the boat.

On that day we were very near the ocean's version of "the other side of the black stump" and had been fishing with various degrees of success for about six weeks since our last port stop.

The helicopters involved in this type of operation are almost exclusively Hughes 500Cs (Bell 47s, the previous mainstays for this kind of work, are now rare on boats in that part of the world). The machines are generally cheap, stripped-down ex-military machines with C18 Allison turbines, datalink GPS and a vessel-specific avionics fit-out. They are also fitted with fixed floats in place of the normal skid configuration. The result is a lightweight, fast and highly manoeuvrable aircraft with a small deck footprint – qualities that are almost essential when you are operating from small, cluttered helidecks in sometimes-difficult conditions.

On these tuna scouting flights the pilot's job is to follow the observer's instructions. Sometimes conflict arises as a result of language difficulties and unreasonable demands. On this particular day I had already flown two flights when at around four o'clock I was asked to make another, this time with the vessel's captain onboard as the observer. He was a hard man who didn't take kindly to any decisions that didn't accommodate his wishes.

I took off and as I climbed away from the boat I noticed a very slight high-frequency vibration coming though the machine, almost imperceptible, but worrying all the same. As every helicopter pilot will tell you, vibrations are a matter of course and an accepted part of being in a helicopter. This time, however, I had a gut feeling that something was wrong. The more I climbed the more alarmed I became. I changed airspeed, power settings and made some slight turns in an effort to reduce or identify the source of the vibration.

By this stage, the ship's captain was shouting at me to get on with it. But I was not going to put up with this vibration for the next two hours – particularly not at 1,500 or 2,000 ft above the sea, sixty-odd miles from the nearest safe landing spot. I shook my head and turned the machine back to the boat with a clearly unhappy passenger.

Upon landing, the captain got out and the ship's aircraft engineer – who hailed from the Philippines – got in. With the engine still running, I tried to explain the problem over the intercom. This plan was thwarted by the engineer's limited English and my non-existent Filipino, so he strapped in and we flew a close circuit before landing and shutdown.

He indicated he understood what I was talking about and we spent the next half hour or so inspecting the machine. We found nothing out of place so I restarted the helicopter and we sat at ground and then flight idle for around 10 minutes without any obvious causal areas showing up. The engineer then jumped in and we took off for a short flight to see if the problem was still there.

Jolt from behind: At around 200-300ft and at about 60 kt the vibration suddenly returned and as the mechanic started signalling me to turn back, I suddenly felt a jolt from behind and the machine immediately rolled hard over to the right with the cyclic flailing later-ally, literally to the stops.

In the blink of an eye we were facing back the way we had come (the bank angle must have exceeded 90 degrees) and the nose was pointing almost vertically down at the water. Through the cyclic I could feel a grinding motion, as if a bearing was self-destructing.

My first thought was that I had lost something in the tail rotor so I dumped the collective and stomped on the left pedal. Either through my efforts, or by chance, the aircraft stopped turning. We were still looking directly down at the ocean and the airspeed was increasing rapidly.

My next thought was, "OK, if I can't pull aft cyclic and get the nose up we are going to die". I pulled aft as hard as I could, and yanked the collective "through the roof". As I braced my feet on the pedals, we hit the water.

When I regained my senses I was underwater and upside down in the cockpit. Although I'm sure that only seconds had passed, I remember thinking, "I'm going to drown because I must have broken something hitting the water with such force". While waiting for the pain I undid my harness and swam out my side of the aircraft (we flew without doors). I figured it would be better to get to the surface and then go back for the mechanic if he hadn't made it out. On surfacing I saw him. His arms were wrapped around the now vertical and ripped starboard float. This was now the only flotation attached to the machine as the port skid had detached and was floating away on the breeze. The net boat was alongside within five minutes, and towed the wreck back to the mother ship where it was lifted inboard with the main derrick.

Our inspection revealed the swash plates were apart from each other by 3-6 inches (depending on collective position). The bearing inside was totally destroyed, the carrier was mangled and various ball bearings were missing or mashed in the rubber boot above the rotating swash plate.

Only two main rotor blades were left and these were wrapped neatly around the rotor head. The other two departed when they hit the rear cabin and tail boom. The tail boom and tail rotor blades were bent. I think of myself as luck to have escaped with a broken rib, broken toe, a large cut on my head and a gash on my foot. It could have been a lot worse.

I was very lucky that my head didn't hit any part of the airframe during impact and I now wear a helmet. I was also fortunate that I was not wearing the provided life jacket, a fixed buoyancy type that would have restricted egress. I now use a manually inflatable jacket and spend time "dry practising" egress drills and emergency procedures in the aircraft I fly.

Lessons Learned:

In this case, the mechanical problems appear to have been the result of a failure of the swashplate bearing. The swashplate is one item called up in the pilot's pre-flight inspection procedure for the Hughes 500 model helicopter. Had this item been adequately inspected before the flight, it is quite likely that indications of impending failure such as looseness, discolouring, wear and debris may have been found.

The pilot's decision to return to the ship following onset of the vibration was correct. Unfortunately, from then on, he was at the mercy of the engineer's advice and competing commercial pressures. One doesn't know the nature of the inspection performed by the engineer following the first event, but it is reasonable to assume that the engineer was unsure of the cause and wanted a better understanding of the problem which could only be gained by a check flight.

With the benefit of hindsight and a bit more experience, the pilot may well have insisted that more possibilities be eliminated before committing himself to a flight in a suspect aircraft.

The sudden onset of vibrations in a helicopter is invariably the forerunner of a serious problem – that's why most flight manuals advise you to land immediately if an unexpected vibration occurs.

While both occupants survived the ditching relatively unscathed, it does not appear that either of them would have completed helicopter emergency underwater (HUET) training. There is also no reference to any emergency equipment such as lifejackets or underwater breathing equipment. In operations where 100 per cent of flight time is overwater, it would make sense to give crews the best chance possible by giving appropriate training and ensuring that right equipment is available should it be needed.

Mal Walker
CASA Flying Operations Inspector

NOTES:

DEEP TROUBLE

CASA

Bell JetRanger, Commercial, Phase: Landing, Name Withheld, Oct 2004

I was settling in nicely to my first year as a commercial helicopter pilot. I had a dream job in Queensland's tropical north, operating a Bell JetRanger on tourism and general charter. The company's main office was in Townsville, well to the south of my operational base in Mission Beach. For the most part, if the bookings kept coming in and the aircraft was flying, they left me alone to just make the money. My little patch of the aviation world covered Dunk Island and the beach resorts of the area. At around 350 hours total helicopter time I was just into that dangerously comfortable zone, where the nervousness goes and you begin to enjoy the flying.

When the office called and told me the aircraft needed routine maintenance and would be grounded for a day, I hardly questioned the reasons. Instead, I planned for the rare bonus of an extra day off. Alarm bells should have been ringing when I met the engineer out at the paddock where the aircraft sat. He was probably the fourth or fifth guy to work on the machine in my short time with the company. Each time it was a different maintenance organisation, different people, different ways.

This time the help arrived in a very old, very beat up Holden. We were a good two hours drive from any major town, so who knows where this guy came from. He proceeded to pull out tools and ladders and set about tearing into the fragile little Allison Gas Turbine that kept me in the air. Once he was settled into his task, I made excuses, and headed off to enjoy the day off. Looking back on it now I should have realised that changing a compressor on a gas turbine was a fairly serious job, probably one that was not best done out of the back of a car, in the middle of a paddock by someone you have never seen or heard of before.

I should have asked a thousand questions, helped him with every task possible and grilled my company about it. But I didn't. I went swimming instead! Three flight hours later, the Allison was purring along nicely. I'd done the post maintenance checks with the engineer after his work, I'd done a test flight, and I'd even clocked up a few hours of scenics and local tours. I was back in my comfort zone, and all was back as it should be. On this day, I had my four tourists on board from Dunk Island, with snorkelling gear and a picnic in the back, and we were off for a day of reef adventure to Pelorus Island. I had done the trip plenty of times before and I filled the half-hour, over-water flight with the most informative commentary I could muster. With Pelorus Island looming up in the front window, I cancelled SAR on the radio. Coverage was poor any lower, so I always got it out of the way at top of decent.

I lowered the collective and began my cruise decent from 1,000 ft. It should have taken us all the way to our beach landing but suddenly an airframe vibration started and quickly built to a violent shudder. It peaked in a violent explosion, which was immediately followed by every warning bell and light simultaneously fighting for my attention.

The alarms were screaming at me as I blurted out a Mayday and jammed the collective lever to the floor. The altimeter was unwinding at a frightening pace as the horizon began to rise up in the windshield. I think there was screaming from my passengers but I busied myself with arming the floats then pulling some kind of flare. All too soon, we hit the water. Water rushed in from broken chin bubbles, as the machine rolled. In an instant we were upside down, it was dark and my senses were filled with the sounds and pressures of lots of water finding its place. I managed to get my passengers out of the machine just in time to watch it sink to the depths below me. We popped our life vests, and spent an uncomfortable hour and a half bobbing in the ocean before finally being plucked to safety by a fishing trawler.

In the many years that have passed since the ditching, I have had plenty of time to think about what I could have done differently. Some of my most enjoyable times in professional aviation have been spent in small helicopter operations where there is the opportunity to get to know, understand and appreciate the engineers who work on the machines. These people are pivotal to the enjoyment of your work and your personal longevity. Get to know them, and take an interest in their work. But above all, take the time to understand what is being done to the aircraft you fl y, and if it doesn't make sense or look right to you, question it. No one can ever ridicule you if you inquire politely and seek knowledge for the right reasons.

Second, as to the way I fly nowadays, I can't see the point in wasting all that sky above me. In tiger country like a heavy forest or unforgiving ocean, I never cruise along at a thousand feet anymore. Had I been higher on the day of the accident, I would have effected a much more accurate Mayday and dramatically reduced the time I spent looking for sharks circling me in the water. And I certainly don't cancel SAR when I'm still in the air and over water. There was a phone on that island and I should have used it after every flight.

Finally, if it all does go wrong, I make sure I have the best chance of getting found. I don't rely on the emergency equipment in the machine, because you may just find all that good gear on the way to the bottom of the ocean.

As a minimum, I carry a small signal mirror on me, a charged up mobile phone, and sometimes much more depending on the flight. I was lucky to come out of a bad situation early in my career. It allowed me to learn and adapt. I hope I've become a better pilot for it.

Lessons Learned:

The pilot's analysis of this accident is well considered and contains valuable lessons. I congratulate him for sharing the story so that others may learn from his experience. This accident occurred when our industry was in a boom phase. Many new operators arrived on the scene and, despite best intentions, some did not have the proper systems in place to adequately manage their safety responsibilities. Since then, a bad economic recession, several cyclones and a drop off in government work has sent many operators to the wall in northern Australia. There has also been an uptake in safety management programs and the benefits to the industry have been considerable. The pilot's experience illustrates the value of maintaining SAR coverage from lift-off to shutdown. This point is highlighted by ATSB statistics showing that a large percentage of helicopter accidents occur while hovering on the way to and from the landing area.

It is better not to cancel flight following until you have landed and fully lowered the collective after a flight. This is especially important approaching a remote area over water, where locals might not see you ditch. I am familiar with this accident and it's worth noting that the pilot managed the ditching, and the events following it extremely well; his passengers can be thankful for his calmness and quick thinking under pressure. Following impact, the floatation gear did not work as designed and this, coupled with a choppy sea, meant the Bell 206 quickly sank, trapping a British and an American middle-aged couple in the cabin. The pilot had to remove his life jacket to dive several times to retrieve the distressed passengers. The next problem he faced was a passenger who panicked and tried to convince everyone they were going to drown. Two of the other passengers, sensing rescue was some time away, calmly reinforced the pilot's reassurances to the distressed man that they would be okay: they had life jackets (except the pilot who lost his while diving for the passengers) and by huddling together and following the pilot's instructions, things would work in their favour. His leadership and command of the situation deserves high praise.

Help your engineer: But Murphy with all his laws does not give up easily: when the trawler came alongside the relieved survivors, the over-eager deck hands caused some injuries when they brought the passengers aboard. No one was injured in the aircraft crash but some of the passengers were hurt when the well-meaning crew hauled them over the rail.

So what can we learn from this accident?

- It's good manners and cheap life insurance to hang around and help your engineer if you're in a remote area. In this way, you can also respectfully monitor work in progress. A new engineer, or one who doesn't appear to have the correct kit, should signal possible problems.
- Do not cancel your SARWATCH until you finally put the collective all the way down. There is rarely a need to cancel flight following in the air, even in remote areas
- This story underlines the value of helicopter underwater escape training (HUET) gives pilots the skills to survive a ditching and provide assistance to their passengers.
- Read up on leadership in a survival situation: your demeanour and humour will make all the difference.
- Finally, ask our boatie mates to be gentle with our passengers. It is easy to crack ribs when hauling someone onto a rescue craft.

Rob Rich
President
Helicopter Association of Australasia

<u>NOTES:</u>

A ZIP TIE
CALLBACK

R22, Training, Phase: Takeoff, Name Withheld, Aug 2008

I pulled the helicopter on a platform out of the hangar after a very thorough preflight done in detail to show a potential student how to correctly do it.

The start was uneventful and I pulled the collective to a manifold pressure of 21 to get the aircraft light on the skids. Then as I increased slowly to lift off I felt a bump and the helicopter took off suddenly, slightly tilted to the left. I was immediately over the edge of the landing platform, one skid over the ground and the other over the platform that is about 3 ft high. I could not control the collective smoothly to get back over the platform and land safely, so I moved forward into the parking lot. Then I lowered the collective and with the same bump the manifold pressure dropped to 19.

The helicopter went down fast towards the ground and I pulled back again to avoid a very hard landing and worse. The helicopter jumped up about 20 ft. I switched hands to check the control friction (I was flying from the left seat to allow the potential student to 'enjoy' flying from the captains seat) and found it normally unlocked.

Through this struggle I brought the helicopter into a more open space (no airplanes), I lowered the collective again and had the same drop but I was ready and as soon as it started to drop I pulled up and caught the fall at about 5 ft above the surface.

Before it started climbing I lowered the collective with the same 'bump' and started sinking. At that point I raised the collective and immediately lowered it to settle amazingly well on the ground. It was not a perfect landing but it was not even a hard landing.

When I started the engine my intent was to bring the helicopter down from the platform to reposition in the parking lot and then call tower and get ready to fly off. After I shut down, the tower asked for explanations which I provided, similar to this report but less detailed. The controller told me he was very concerned, even more so because I was not talking to him, which would have not helped, could even have distracted me.

After this landing we found a broken zip tie sticking out of the left collective boot. After removing the boot completely the rest of the zip tie was found attached to the clip holding the left collective in place (this is an early model with a poorly designed removable collective, difficult to put back in).

The presence of the zip tie there remains completely unexplained as it was not there before the aircraft went into a major overhaul. I have no doubt about this issue since I removed the left collective several times and I never had anything attached to the clip.

After this unpleasant adventure I removed the landing light that needed changing and found inside the bottom of the aircraft 2 washers, 2 cut zip ties, a clip, and a few other miscellaneous small items. I removed all this hardware and performed the landing light change.

Overall it was an experience that ended up well by some luck and maybe some skills on my part. I carefully checked the left collective before removing it and after placing it back. The zip tie was sliding freely along the clip and could easily get caught between the collective and the frame, rubbing against the frame of the aircraft in the opening where the collective goes through the firewall.

My attempt at controlling the aircraft ended up cutting it against the sharp edge and most likely explains that I was able to land in a normal manner after all these out of control movements, more luck than skills I guess.

Lessons Learned:

The intent of this report is to insist once more that after a major overhaul there are gremlins. The aircraft had been flown about 7-8 hrs before this incident but the zip tie could move freely along the clip and could get caught at any time or never with luck.

NOTES:

FUELLED BY LIMITED OPTIONS
CALLBACK

Agusta Unidentified, Passenger, Phase: Cruise, Name Withheld, Apr 2014

Early morning departure, enroute to oil platform with passengers on board. We had enough fuel to complete the flight with required reserves, taking into consideration the forecast headwind. We noted alternate fuel stops for the forecasted wind conditions, and during the flight we determined that we had enough fuel to complete the flight without requiring an additional stop.

During cruise we ran a manual fuel burn calculation to further verify that an additional stop was not necessary. We called the platform 20 minutes out requesting fuel and we were assured fuel is available and to call 5 minutes out for a green deck as standard procedure.

At that point we knew we had two alternate fuel stops within range, but determined to continue to our destination since at that time we were not advised of any expected delay or deck blockages.

About 4 minutes after our 20 minute call to the platform we were advised by the Helicopter Landing Officer (HLO) that an approaching aircraft from another company required refuelling. We notified the HLO of our fuel status and that we did not have time to loiter. This message was acknowledged by both the HLO and landing aircraft. We anticipated a slight delay due to the other aircraft on deck, who also had minimum fuel, so we slowed our airspeed to decrease our fuel burn.

About 10 minutes from reaching the platform the HLO notified us that the fuel system was not working due to an electrical problem with no known timeframe for the repair. We confirmed with the other helicopter that he was now on our destination and did not have enough fuel to move and allow us to land. As soon as we heard this information we started scanning for alternate landing structures that would safely support our aircraft. Prior to reaching our reserve fuel we saw a vessel to the west and immediately deviated towards it.

We noticed that the vessel did have a large enough heli-deck and proper weight rating so we contacted them on Marine 16 and requested a green deck while orbiting. After four orbits the Captain of the vessel confirmed we had a green deck, so we proceeded inbound and safely. The passengers were then boated from the vessel to the platform while we waited for fuel to be delivered.

Fuel was delivered to the vessel and [12 hours later] a mechanic was hoisted onto the heli-deck. The fuel was sumped multiple times and the transfer vessels were assured clean before fuel was transferred to the helicopter. We then sumped the aircraft the next morning after the fuel had settled to ensure we had a clean sample and then departed the vessel.

Lessons Learned:

Destination being frequently used by a non-contract aircraft as a fuel stop and us not being notified. Destination allowing an alternate aircraft to land within +/- 30 minutes of scheduled crew change flight, even though our ETA was told to the customer. Fuel system at destination having known electrical glitches and not being passed along to inbound aircraft.

NOTES:

FERRYING A DAMAGED R22
CALLBACK

R22, Ferry, Phase: Takeoff, Name Withheld, May 2004

I work at a helicopter company as a pilot/instructor and mechanic. A renter pilot had a hard landing at a nearby gravel pit and he called me to come and look at the helicopter.

I flew a different helicopter to the site and upon arrival, I found that the helicopter had a bent vert brace to which the landing gear is bolted. Pilot informed me that he had called the FAA and they told him to call back after a mechanic looked at the aircraft and if it had substantial damage.

I inspected the helicopter thoroughly and found no other damage. I determined that the damaged part was not essential for flight and would not affect the flight characteristics of the helicopter. I had also determined that the helicopter could be safely flown back to our hangar which was a 5 min flight.

At that time there was a severe thunderstorm approaching and because for the past few days we have had severe thunderstorms with high winds and hail, I feared for further damage to the helicopter. With it being Sunday, and nobody at the gravel pit for us to load the helicopter on a trailer, the approaching storm, dealing with an upset pilot, and considering the damage not to be substantial, I decided to fly the helicopter back to our hangar.

Later that afternoon, I received a phone call from the FAA and I told him about the damage and that I flew the helicopter back to our hangar. The inspector told me that there might be an issue as regards a ferry permit and he would be out to inspect the helicopter the next day. I told the inspector that due to the circumstances at the time, I had overlooked the fact that I might need a ferry permit.

The next day the inspector looked at the aircraft and said that in his opinion, the bent part would be considered substantial damage and he would report this as an accident rather than an incident. The inspector also said that he would have issued a ferry permit had I called him.

Lessons Learned:

I feel the things that contributed to this aircraft being flown without a ferry permit are:

1. The approaching storm with possible further damage to aircraft.
2. The vagueness of the conversation between the pilot and FAA, leaving the decision of substantial damage to me (the mechanic).
3. Dealing with upset pilot (a distraction).
4. Obtaining a ferry permit is not an everyday occurrence, so therefore easily overlooked.

Callback Comment: The reporter said the damage was one bent vertical brace to which the leading gear is bolted. The reporter stated it was minor damage and believed the inspector was mainly upset that no phone call was received asking for a ferry permit. The reporter said there was a short time to move the aircraft as severe thunderstorms with hail and high winds were approaching the area and may damage the aircraft. The reporter stated finding a telephone in a gravel pit on Sunday was impossible and he elected to fly the machine out immediately.

NOTES:

CHAPTER 9

AIRWORTHINESS & MAINTENANCE

"This wonderful flying machine [the helicopter] is, in some ways, an ultimate in airplanes."
Thomas Alva Edison, 1908

DIMINISHED ENGINE POWER
CALLBACK

R44, Private, Phase: Takeoff, Name Withheld, Jul 2013

I flew a Robinson R44 on a Part 91 personal flight from the helicopter pad with two passengers. Weather on the date of the flight included a local altimeter setting of 29.96, density altitude of 1,800 FT and light winds out of the Southeast. During run-up, the helicopter's OAT gauge read 95 degrees Fahrenheit on the concrete helipad, and approximately 89 degrees over the grass. Nearby flags indicated winds directly out of the east at approximately 8-12 MPH. The helicopter's total weight on the incident flight was approximately 2,175 LBS. The R44's gross weight is 2,400 LBS. At the given weight and temperature, the R44 has a demonstrated ability to maintain a zero wind OGE (Out of Ground Effect) hover at approximately 3,000 ft AGL. In other words, the helicopter was operated well inside its published operating envelope.

I performed a normal run up and all checklist items appeared normal. Upon takeoff, the helicopter developed insufficient climb power. I dropped off a passenger, and thinking the helicopter would now climb easily, entered a second maximum performance takeoff.

The helicopter again developed insufficient power, but the engine did not miss, surge, backfire, or give any other indication of partial lack of power.

Seconds into the flight, rotor RPM decayed and I entered an autorotation and made an emergency landing in a private front yard, having narrowly missed several houses and other obstacles.

After the remaining passenger departed, the engine suddenly surged, then appeared to behave normally. As bystanders approached the aircraft from all directions, including the direction of the tail rotor, and fearing for their safety, I made the immediate decision to takeoff again. The helicopter performed perfectly on each subsequent flight.

The cause of the loss of power has not yet been determined.

NOTES:

DEVELOPED COMPRESSOR STALL
CALLBACK

AS350, Utility, Phase: Cruise, Name Withheld, Jul 2010

I was piloting a law enforcement Eurocopter AS350BA responding to assist looking for a missing person. I was flying with an experienced Tactical Flight Officer (TFO) who is rated in airplanes only. When we arrived over the incident location we were at 500 ft AGL and approximately 100 KTS. I lowered the collective to decelerate and also began a shallow left hand turn.

I noticed the Ng passing below 90% but the bleed valve light was not illuminated, as it usually is below 96%. At the same time I heard the sound of compressor stalls from above and behind me. The stalls were continuous and were later reported being heard by deputies on the ground working the same call. I raised the collective per the emergency procedure, but the compressor stalls continued. I continued to manipulate the collective to try and find a position where the condition would stop, but was unable to find one.

At that point I noticed the rotor RPM was lowering to the bottom of the yellow arc and I heard the low rotor RPM horn sound. I fully lowered the collective, made a 90 degree turn to the right towards a four lane road and broadcast a 'Mayday' over the police radio.

My TFO assisted with transmitting our location and asking for additional units while looking out for obstacles and pointing out possible alternative landing locations. When I started the flare to land I found the aircraft to be unstable in the yaw axis as the compressor stalls continued so I elected to do a run-on landing. The engine was still running upon landing and I shut it down immediately with my TFO reading off the checklist to confirm all systems were shut down.

There was no damage to the aircraft or property on the ground. The entire event, from first compressor stall to touchdown was approximately 20 seconds.

NOTES:

ENGINE CHIP
CALLBACK

AS350, Ambulance, Phase: Cruise, Oct 2010

Following aircraft refuelling, the crew and aircraft were heading back to home base. Approximately 12 minutes into the flight leg, the aircraft's "Eng Chip" caution warning panel light illuminated.

I immediately checked my engine instruments (all were within normal operating range) and reversed course to return to a large open field we had just flown over. I then notified Departure Control and my medical crew of the situation and my intentions to turn the aircraft around. Departure authorized my course reversal.

Upon arrival at the open field I informed Departure and the medical crew of my landing intentions and began an expedited descent to begin my approach to land. At this stage, all of the aircraft's instruments were still within operating limits and there were no warning lights [other than] the "Eng Chip" light.

At approximately 200 ft AGL we heard a loud bang, that seemed to have come from the engine area, and experienced a sharp left yaw in the aircraft. Upon hearing the bang and feeling the aircraft yaw, I immediately entered an auto-rotational profile toward the open field, as I fully expected the engine to quit running. It was then that I also noticed a strong "burning" smell.

At approximately 65 ft AGL I began a flare to retard my airspeed and my Flight Paramedic informed me that, "We are on fire".

I terminated my expedited approach to the ground and immediately executed on-board aircraft fire procedures by shutting both aircraft fuel levers and motoring the starter with the "Crank" button, although I never got a "Fire" light on the caution warning panel.

In the meantime, the Flight Paramedic had deplaned with the aircraft's fire extinguisher and was standing on the port side of the aircraft ready to fight the fire. I advised him not to use the extinguisher until we were sure of the location of the flames. Fortunately there were no flames and the smoke coming from the aircraft's exhaust was starting to dissipate.

The Base Mechanic, Lead Mechanic, Dispatch Center and ATC were all notified of the successful emergency landing without injury or aircraft damage. The Base Medical Manager arrived at the scene to pick up the medical crew and transport them back to base. I remained with the aircraft and the Base Mechanic arrived at the scene. The Mechanic and I stayed on scene to inspect and secure the aircraft and [later] departed for base. Arrangements were made by the Base Mechanic to load the helicopter on a flatbed truck and transport back to base.

NOTES:

A DIMLY LIT DENT
CALLBACK

BK-117, Medical, Phase: Pre-flight, Name Withheld, Jan 2012

I conducted a pre-flight in the morning upon reporting for my shift and no aircraft deficiencies were noted.

I received a request for a patient pick-up at one hospital for transport to another hospital. I conducted a pre-start walk around with no deficiencies noted and flew to the [first] hospital.

After shutdown, I conducted a post-flight walk around with no deficiencies noted. I loaded the patient, conducted another pre-start walk around with no problems noted and flew to the receiving hospital. After shutdown at the receiving hospital, I conducted a post-flight walk around and serviced the helicopter.

During my pre-start walk around prior to returning to base, I noted a shadow on one of the tail rotor blades that "didn't look right." Upon close-up visual examination and moving the blades to just the right position in relation to the sun, I noted a dent in the blade that, when touched, felt to me like a de-lamination in the fibreglass outer shell from the inner foam core of the blade itself.

Lessons Learned:

If something does not look right, feel right or smell right; stop, assess the situation and determine a course of action.

On this day, I personally looked at the tail rotor on five separate occasions and did not notice any deficiencies.

During the execution of the sixth walk around of the day, the sun was at just the right angle to create a shadow that allowed me to discover what could have been a catastrophic flaw in the tail rotor.

Interestingly, one hour later, it took three of us to find the dent again since the sun had moved enough to create a different lighting angle on the blade.

<u>NOTES:</u>

TOO MUCH IRON
CALLBACK

AS350B2, Commercial, Phase: Parked, Name Withheld, Aug 2016

This helicopter is used as an electronic news gathering ship. Every 100 hours an engine oil sample is taken for analysis. During the past four samples, the iron content has spiked.

After the initial report was received the manufactures tech rep was advised and sent the report. His recommendation was to decrease the oil change interval to 200 hours from 300 hours to lessen the amount of contamination passing through the engine. At that time, the sample level (4.0 parts per million) was less than the threshold level of 7.5 ppm requiring a maintenance action. The next sample was at 8.0 ppm and the maintenance manual calls for three daily samples to establish the accumulation rate.

This requirement was added to the aircraft status sheet which is reviewed by the flight crew's daily and the director of maintenance (DOM) was advised. The three samples were completely ignored and the aircraft continued in service.

At the next 100 hour, the soap sample results had the iron content at 8.7 ppm. The oil was changed but no other actions were taken. I was away from the shop for two weeks and upon my return, no other maintenance or sampling had taken place. I asked the DOM to have a tech take a sample and it was sent out. The sample returned at 91 hours on the fresh oil with an iron level of 8.2 ppm.

The manufactures rep was again contacted and he stated that if any fuzz or sludge was present on the engine chip plugs, the engine should be taken out of service. This was communicated verbally and by email to the DOM. A shop technician stated that there had been sludge present at the last chip plug inspection and this was ignored by the DOM.

The wear within the engine had increased. Again the proper procedure was to initiate a three consecutive sampling program but this was not done per the DOM. Instead the oil was changed. At the next sample with fresh oil, the iron level was 9.8 ppm with 81 hours of operating time. I advised the DOM to take the helicopter out of service but he stated to keep it in service and do another oil change and sample at the next 100 hour.

The engine has a total time of 3400 hours since new with a TBO of 3600 hours. The reason given by the DOM was a replacement engine was not available for 30 days.

The engine removal level is anything greater or equal to 15 ppm, or an accumulation rate of greater than .5 mg / hr. These engines have a documented failure rate of certain accessory gears and the tech rep stated that is probably where the iron is being generated from.

The prudent action would be to remove the engine from service prior to TBO and have the problem investigated. Due to management's desire to satisfy contract obligations with the TV station, they have elected to keep the aircraft in service. At the last inspection, there were numerous cracks in the engine firewall which required replacement indicating a possible engine vibration problem. The information and signs are in place of a future failure and management continues to ignore the warning signs.

Lessons Learned:

This is a precursor to an accident and it will be obvious to any investigating authority there was a known problem and management chose to do nothing about it.

NOTES:

CONDITIONS FORCE ERRORS
CALLBACK

AS350B3, Ambulance, Phase: Parked, Name Withheld, Jun 2015

Hangar Mechanic 1

I work in a 145 Repair station as a hangar mechanic, and I was asked to go to assist in a swashplate change on an AS350B3 that was stuck on the hospital landing pad due to a swashplate shim protruding from the center of the swashplate. The repair took 11 days. We had been working very long hours (10-14hrs daily) in very hot weather all outside on top of the Hospital pad where the helo was stranded. During the swashplate change after disassembly of the main gear box we noticed that the bolt holes had corrosion build up in them and the ring gear had also been corroding due to bad sealant on the bolt heads, so we ended up removing the entire assembly and changed out the epicyclic and housing at the same time. During the reassembly of the housing and epicyclic, (we had been working for 14 hours that day) we decided to call it and to get some rest. I left to go to my hotel and came back in the morning to find that the epicyclic had already been installed into the housing, however, during my RII of the installation I noticed that the gearbox and epicyclic had been coated with Turco instead of mineral oil which is what is called out for. I notified the lead and he said they had installed it after I left. So they came up to the pad to confirm and ended up flushing the system out before we installed the rest of the mast and components. The lead mechanic was instructing and helping to assemble the mast and swashplate assembly (the Maintenance Manual (MM) reference was printed out and we were going step by step). We had to take breaks often due to how hot it would get on top of the hospital but the job got finished.

[Several months later] the helicopter was trucked in for a tail rotor strike. During a brief incoming inspection and cleaning another mechanic noticed that the dust cover and or boot that covers the top of the swashplate was missing.

During our maintenance ops, there were many people constantly calling to figure out when we would be done which caused many distractions, so less or no phone calls during maintenance ops on an Aircraft on Ground (AOG) aircraft would be immensely helpful. If the aircraft is stuck on the helipad or in an area where it cannot be moved to a hangar, putting up a temp shelter (roof tent) or something like that would help immensely with less fatigue and possible heat stroke, which in turn leads to more mistakes and less work time due to the time spent trying to cool down and regulate body temp and sunburn. No pilot interaction during these maintenance ops would also help a lot, it makes our jobs as mechanics a lot harder when we not only have supervisors and their bosses phoning us all the time to then have the pilot ask how much longer.

Lead Mechanic

This aircraft was removed from service and placed under unscheduled maintenance for removable shims extruding from the swashplate on top of the hospital helipad. A [contract] was submitted for outside help as the maintenance crew on staff had not performed this job to a level of comfort. Maintenance support arrived for assistance. Upon removal of the mast assembly, it was noted that the epicyclic gearbox and sun gear assembly were corroded beyond limits, further increasing the workload on the [roof] of the hospital in the middle of summer, requiring a full transmission replacement. Initially, we were told that removing the aircraft from the hospital by crane was not the best option at the time, as we were only to be performing a couple days' worth of work, but when the corrosion was found, it would have taken the same amount of time to reinstall everything needed to crane the aircraft as to put it back together to return to service. The removal, disassembly, reassembly, and return to service took us 10 days of all mechanics working over 10 hours per day. During this time I worked on and off with the mechanics performing the maintenance, as I am the lead mechanic for the program and had other priorities. During this 10 day event I called a safety stand down day on day 7 as the maintenance crew was showing signs of heavy fatigue, both mentally and physically. The installation, ground runs, and check flights were completed and the aircraft was returned to revenue service. While I was not the primary mechanic on this job, I feel it necessary to submit this as I did help and did not notice the issue.

[Over 2 months later] this aircraft was involved in a tail rotor strike at a scene call. This aircraft was loaded on a truck and shipped to a repair station for maintenance required for the tail rotor strike inspection. During acceptance inspection, it was noted that the swashplate appeared to be missing the dust cover on top of the swash-plate. I was notified by my supervisor, and began researching this issue. After consulting the removal and installation part numbers, it became apparent that the installed part number was not effective for this aircraft. After consulting the maintenance staff involved, and looking at the Illustrated Parts Catalog (IPC) for this install, we could not find the correct swashplate in the IPC, and the only part number shown was the one that was installed. The repair station has the aircraft at this time, and will make the repairs as necessary. This event was an eye opener to all of us as we look back to all of the long hours that were put in to get this aircraft back in service, and that sometimes we need to take a step back and look at ourselves before we look at the work ahead of us.

In my opinion, there were multiple factors that lead to this incident. Maintenance staff was pressured, both personally and by the customer, to provide an expedited service that could not be met. Maintenance pressures were also felt from the heat during this time of the year, on top of an aluminium helipad, with no shade available at the aircraft. The airframe IPC played a large role in the incorrect ordering of parts, as the IPC is somewhat vague when it comes to ordering parts for aircraft with different modifications installed (i.e., dual hydraulics in this case). I believe this incident could have been avoided if the aircraft had been removed from the hospital from the very beginning, removing the pressures from the hospital on a direct level.

Hangar Mechanic 2

I was notified that an aircraft had been sent by ground transport to the Repair Station due to a Tail Rotor strike. I was made aware that during the receiving inspection into the repair station that it was discovered that the Main Rotor Swashplate assembly was missing the top dust cover (boot). Later during this day it was also brought to my attention by the Lead Mechanic that after discussing this situation with Mechanic that was currently on duty that the Swashplate that is currently installed may be the incorrect assembly.

Due to a previous recent event, the swashplate had been replaced. The replacement was due to a peel shim that was found during a preflight check that was partially ejected from the top of the uniball. The work for this replacement was [accomplished several months ago]. While attempting to order the replacement swashplate I was unable to find the replacement part number for this assembly in the online AS350B3 Illustrated Parts Catalog. I notified the lead mechanic to inform him that I was unable to find this part number. After some discussion we both agreed that the Airbus Helicopter Technical Representative should be called for support.

I immediately called the Tech. Rep. and explained my situation and that this was to be installed on an AS350B3 with Dual Hydraulics. I was told that I would not find that part number in the AS350B3 IPC and that I would need the part numbers found on the Component Cards for the Rotating and Non-rotating Swashplates (sub-assembly part numbers) and that he would send me a sheet that would give a breakdown of these part numbers that would provide the Swashplate Assembly part number (Next Higher Assembly) that was needed. This Swashplate was ordered and properly received into the base.

Before the work for the replacement of this Swashplate began, I voiced concerns with the Lead Mechanic that although I have performed this maintenance tasking before for a swashplate replacement on an AS350B3, it had been maybe as long as 6 years since the last time I had done this and I had not done this task on an AS350B3 with Dual Hydraulics. This coupled with the fact that this was a brand new base and that this maintenance was to be performed on the roof of the hospital with temperatures that were to be in the high 90s were also concerns that I voiced. The lead mechanic agreed that we would need additional support. He made arrangements to have a mechanic experienced at this procedure sent from another Repair Station and requested that I was to perform this as OJT that was to be documented in the OJT book provided to the field mechanics. Another experienced Mechanic was also sent for additional maintenance support.

Although another mechanic signed off this installation, the three mechanics, myself included, involved in this installation shared equally the responsibility of this installation. I provided the RII and at that time I should have reviewed the installation procedure again with the two mechanics. I believe that each one of us were under the impression (although now knowing it to be wrong) that because the assembly that came off did not require the boot that this was still the case with new assembly that went on.

Lessons Learned:

I have reviewed the procedure in the AS350B3 Aircraft
Maintenance Manual (AMM) for the installation of the swashplate
onto the Mast Assembly again and it does not distinguish between
the different assembly types with either the required dust boot or a
rain guard as is found on the AS350B3's with dual hydraulics. The
procedure does call out for the installation of the boot and in retro-
spect I feel now that I should have questioned our installation at this
time. There is also nothing that I could find on the hard cards that
distinguish the installation of the rain guard as opposed to the lip ring
used for the installation of the dust boot. I feel the process for determining the part number for the Swashplate assembly is convoluted
and not in good practice. I should have questioned this process at the
time that I did not find the part number in the IPC.

I believe there were more challenges with this swashplate change
and eventually the Main Gearbox than normal in a field environment. This was a brand new base that was not completely equipped
yet to handle this type of heavy maintenance. The fact the work that
was to be completed on the roof of the hospital where the OAT was
95 to 100 degrees F each day and that we were on a metal helipad
was extremely challenging. The teardown of the main gear box and
Mast Assembly was completed in a small room on the roof of the
hospital just outside the elevator with no air conditioning or circula-
tion of air. We worked 10 to 14 hours each day in order to reach our
goals. Word was getting passed on to us about some of the outside
pressures to have the aircraft back in service and that this was a new
base and a new contract that was taken over and that these issues did
not exist with them.

NOTES:

CHAPTER 10

FURTHER READING

IF YOU ENJOYED READING 71 *Lessons From The Sky*, then you might also enjoy the other books in the *Lessons From The Sky* series.

51 Lessons From The Sky (US Air Force)
61 Lessons From The Sky (Military Helicopters)
81 Lessons From The Sky (General Aviation)
101 Lessons From The Sky (Commercial Aviation)
Top Gun Lessons From The Sky (US Navy)

Do you have any lessons you would like to share with Fletcher, and the aviation community? Email them to Fletcher at

fletch@avgasgroup.com

for him to share on his social media accounts, or to include in a future book. Please note that if you do send through any lessons, you are giving us permission to publish those stories.

GLOSSARY

A

Absolute altitude. The actual distance an object is above the ground.

Acceleration. A change in velocity; a change in either speed or direction or both.

Advancing blade. The blade moving in the same direction as the helicopter. In helicopters that have counterclockwise main rotor blade rotation as viewed from above, the advancing blade is in the right half of the rotor disk area during forward movement.

Aerodynamic. Relating to the flow of air around a body and that body's reaction to that flow. An aerodynamic shape is one that allows air to flow smoothly.

Agonic Line. An isogonic line along which there is no magnetic variation.

Air density. The density of the air in terms of mass per unit volume. Dense air has more molecules per unit volume than less dense air. The density of air decreases with altitude above the surface of the earth and with increasing temperature.

Aircraft pitch. The movement of the aircraft about its lateral, or pitch, axis. Movement of the cyclic forward or aft causes the nose of the helicopter to pitch up or down.

Aircraft roll. The movement of the aircraft about its longitudinal axis. Movement of the cyclic right or left causes the helicopter to tilt in that direction.

Airfoil. Any surface designed to obtain a useful reaction of lift, or negative lift, as it moves through the air.

Airframe. The structure of an aircraft. The frame itself, often made of tubes, or the entire fuselage. Typically, in discussions about rotorcraft, the "airframe" refers to the entire helicopter – less the rotors.

Airworthiness Directive. When an unsafe condition exists with an aircraft, the FAA issues an Airworthiness Directive to notify concerned parties of the condition and to describe the appropriate corrective action.

Altimeter. An instrument that indicates flight altitude by sensing pressure changes and displaying altitude in feet or meters.

Angle of attack. The angle between the airfoil's chord line and the relative wind. The AOA and airspeed on an airfoil determine the life and drag of that airfoil.

Angle Of Incidence. Same as angle of attack; sometimes defined as pitch.

Antitorque pedal. The pedal used to control the pitch of the tail rotor or air diffuser in a NOTAR® system.

Antitorque rotor. See tail rotor.

Articulated rotor. A rotor system in which each of the blades is connected to the rotor hub in such a way that it is free to change its pitch angle, and move up and down and fore and aft in its plane of rotation.

ASI. Air Speed Indicator; an instrument used for measuring speed relative to airflow, not relative the ground.

Autopilot. Those units and components that furnish a means of automatically controlling the aircraft.

Autorotation. The condition of flight during which the main rotor is driven only by aerodynamic forces with no power from the engine.

Axis of rotation. The imaginary line about which the rotor rotates. It is represented by a line drawn through the centre of, and perpendicular to, the tip-path plane.

B

Basic empty weight. The weight of the standard helicopter, operational equipment, unusable fuel, and full operating fluids, including full engine oil.

Blade coning. An upward sweep of rotor blades as a result of lift and centrifugal force.

Blade damper. A device attached to the drag hinge to restrain the fore and aft movement of the rotor blade.

Blade feather or feathering. The rotation of the blade around the spanwise (pitch change) axis.

Blade flap. The ability of the rotor blade to move in a vertical direction. Blades may flap independently or in unison.

Blade grip. The part of the hub assembly to which the rotor blades are attached, sometimes referred to as blade forks.

Blade lead or lag. The fore and aft movement of the blade in the plane of rotation. It is sometimes called "hunting" or "dragging."

Blade loading. The load imposed on rotor blades, determined by dividing the total weight of the helicopter by the combined area of all the rotor blades.

Blade root. The part of the blade that attaches to the blade grip.

Blade span. The length of a blade from its tip to its root.

Blade stall. The condition of the rotor blade when it is operating at an angle of attack greater than the maximum angle of lift.

Blade tip. The furthermost part of the blade from the hub of the rotor.

Blade track. The relationship of the blade tips in the plane of rotation. Blades that are in track will move through the same plane of rotation.

Blade tracking. The mechanical procedure used to bring the blades of the rotor into a satisfactory relationship with each other under dynamic conditions so that all blades rotate on a common plane.

Blade twist. The variation in the angle of incidence of a blade between the root and the tip.

Blowback. The tendency of the rotor disk to tilt aft in transition to forward flight as a result of unequal airflow.

C

Calibrated airspeed (CAS). Indicated airspeed of an aircraft, corrected for installation and instrumentation errors.

Camber. The top or bottom curve of an airfoil. If the top and bottom curves are equal (mirror images), the airfoil is symmetrical; if the top has more curvature than the bottom, the airfoil is asymmetrical.

Center of gravity. The theoretical point where the entire weight of the helicopter is considered to be concentrated.

Center of pressure. The point where the resultant of all the aerodynamic forces acting on an airfoil intersects the chord.

Centrifugal clutch. A clutch that engages as a result of an outward force caused by turning motion. A device that mechanically separates, and engages drive transmission between the helicopters engine and combined rotor system. This system is usually automated without any pilot input. Such systems can reduced start-up and shut-down workload requirements for pilots.

Centrifugal force. The apparent force that an object moving along a circular path exerts on the body constraining the object and that acts outwardly away from the center of rotation.

Centripetal force. The force that attracts a body toward its axis of rotation. It is opposite centrifugal force.

Chip detector. A warning device that alerts you to any abnormal wear in a transmission or engine. It consists of a magnetic plug located within the transmission. The magnet attracts any metal particles that have come loose from the bearings or other transmission parts. Most chip detectors have warning lights located on the instrument panel that illuminate when metal particles are picked up.

Chord. An imaginary straight line between the leading and trailing edges of an airfoil section.

Chordwise axis. For semirigid rotors, a term used to describe the flapping or teetering axis of the rotor.

Clutch. A system whether automatic or manual, of engaging the engine drive to the main flight transmission. e.g. Centrifugal Clutch or clutchable V – belts to main rotor gearbox to drive the rotors.

Coaxial rotor. A rotor system utilizing two rotors turning in opposite directions on the same centerline. This system is used to eliminated the need for a tail rotor.

Collective pitch. The collective pitch control, or collective lever, is normally located on the left side of the pilot's seat (with an adjustable friction control to prevent inadvertent movement) changes the pitch (angle of attack) of the main rotor blades collectively to achieve ascent, decent and hover. It is the equal and simultaneous altering the pitch of all main rotor blades; controls vertical flight (height) of a helicopter.

Collective pitch control. The control for changing the pitch of all the rotor blades in the main rotor system equally and simultaneously and, consequently, the amount of lift or thrust being generated.

Coning. See Blade Coning.

Coriolis effect. The tendency of a rotor blade to increase or decrease its velocity in its plane of rotation when the centre of mass moves closer to or farther from the axis of rotation.

Cyclic feathering. The mechanical change of the angle of incidence, or pitch, of individual rotor blades, independent of other blades in the system.

Cyclic pitch. The cyclic control usually located between the pilot's legs and is commonly called the cyclic stick is similar to an aeroplanes ailerons but not only controls rolling left and right to achieve turns, it also moves forward and back to obtain it's directional control. The variation of the pitch of each rotor blade individually during each revolution; controls the horizontal flight direction of a helicopter.

Cyclic pitch control. The control for changing the pitch of each rotor blade individually as it rotates through one cycle to govern the tilt of the rotor disk and, consequently, the direction and velocity of horizontal movement.

D

Delta hinge. A flapping hinge with an axis skewed so that the flapping motion introduces a component of feathering that would result in a restoring force in the flap-wise direction.

Density altitude. Pressure altitude corrected for nonstandard temperature variations.

Deviation. A compass error caused by magnetic disturbances from the electrical and metal components in the aircraft. The correction for this error is displayed on a compass correction card placed near the magnetic compass of the aircraft.

Direct control. The ability to maneuver a helicopter by tilting the rotor disk and changing the pitch of the rotor blades.

Direct shaft turbine. A single-shaft turbine engine in which the compressor and power section are mounted on a common driveshaft.

Disk area. The area swept by the blades of the rotor. It is a circle with its center at the hub and has a radius of one blade length.

Disk loading. The total helicopter weight divided by the rotor disk area.

Dissymmetry of lift. The unequal lift across the rotor disk resulting from the difference in the velocity of air over the advancing blade half and the velocity of air over the retreating blade half of the rotor disk area.

Downwash. The downward rush of air produced by the powered rotor.

Drag. An aerodynamic force on a body acting parallel and opposite to relative wind.

Dual rotor. A rotor system utilizing two main rotors.

Dynamic rollover. The tendency of a helicopter to continue rolling when the critical angle is exceeded, if one gear is on the ground, and the helicopter is pivoting around that point.

F

Feathering. The action that changes the pitch angle of the rotor blades by rotating them around their feathering (spanwise) axis.

Feathering axis. The axis about which the pitch angle of a rotor blade is varied. Sometimes referred to as the spanwise axis.

Feedback. The transmittal of forces, which are initiated by aerodynamic action on rotor blades, to the cockpit controls.

Flapping. The vertical movement of a blade about a flapping hinge.

Flapping hinge. The hinge that permits the rotor blade to flap and thus balance the lift generated by the advancing and retreating blades.

Flare. A maneuver accomplished prior to landing to slow a helicopter.

FPM. Feet per minute describing speed of ascent or descent.

Free turbine. A turboshaft engine with no physical connection between the compressor and power output shaft.

Freewheeling unit. A component of the transmission or power train that automatically disconnects the main rotor from the engine when the engine stops or slows below the equivalent rotor rpm.

Fully articulated rotor system. See articulated rotor system.

G

Gimbal ring. A universal coupling, permitting the swash plate to tilt at any angle relative to the mast.

Governor. An electrical or mechanical device that automatically maintains engine RPM with varying flight loads.

Gravity. See weight.

Gross weight. The sum of the basic empty weight and useful load.

Ground effect. A usually beneficial influence on helicopter performance that occurs while flying close to the ground. It results from a reduction in upwash, downwash, and bladetip vortices, which provide a corresponding decrease in induced drag.

Ground resonance. Self excited vibration occurring whenever the frequency of oscillation of the blades about the lead-lag axis of an articulated rotor becomes the same as the natural frequency of the fuselage.

Gyroscopic procession. An inherent quality of rotating bodies, which causes an applied force to be manifested 90° in the direction of rotation from the point where the force is applied.

H

Hover. A flight maneuver in which the helicopter is maintained in a fixed position above the ground, both vertically and horizontally.

Human factors. The study of how people interact with their environment. In the case of general aviation, it is the study of how pilot performance is influenced by such issues as the design of cockpits, the function of the organs of the body, the effects of emotions, and the interaction and communication with other participants in the aviation community, such as other crew members and air traffic control personnel.

Hunting. Movement of a blade with respect to the other blades in the plane of rotation, sometimes called leading or lagging.

I

In ground effect (IGE) hover. Hovering close to the surface (usually less than one rotor diameter distance above the surface) under the influence of ground effect.

Induced drag. That part of the total drag that is created by the production of lift.

Induced flow. The component of air flowing vertically through the rotor system resulting from the production of lift.

Inertia. The property of matter by which it will remain at rest or in a state of uniform motion in the same direction unless acted upon by some external force.

Isogonic line. Lines on charts that connect points of equal magnetic variation.

K

Knot. A unit of speed equal to one nautical mile per hour.

L

LDMAX. The maximum ratio between total lift (L) and total drag (D). This point provides the best glide speed. Any deviation from the best glide speed increases drag and reduces the distance you can glide.

Lateral vibration. A vibration in which the movement is in a lateral direction, such as imbalance of the main rotor.

Lead and Lag. The fore (lead) and aft (lag) movement of the rotor blade in the plane of rotation.

Licensed empty weight. Basic empty weight not including full engine oil, just undrainable oil.

Lift. One of the four main forces acting on a helicopter. It acts perpendicular to the relative wind.

Load factor. The ratio of a specified load weight to the total weight of the aircraft.

M

Married needles. A term used when two hands of an instrument are superimposed over each other, as on the engine/rotor tachometer.

Mast. The component that supports the main rotor.

Mast bumping. Action of the rotor head striking the mast, occurring on underslung rotors only.

N

Navigational aid (NAVAID). Any visual or electronic device, airborne or on the surface, that provides point-to-point guidance information, or position data, to aircraft in flight.

Night. The time between the end of evening civil twilight and the beginning of morning civil twilight, as published in the American Air Almanac.

Normally aspirated engine. An engine that does not compensate for decreases in atmospheric pressure through turbocharging or other means.

O

One-to-one vibration. A low frequency vibration having one beat per revolution of the rotor. This vibration can be either lateral, vertical, or horizontal.

Out of ground effect (OGE) hover. Hovering a distance greater than one disk diameter above the surface. Because induced drag is greater while hovering out of ground effect, it takes more power to achieve a hover out of ground effect.

P

Parasite drag. The part of total drag created by the form or shape of helicopter parts.

Payload. The term used for the combined weight of passengers, baggage, and cargo.

Pedals. Referred to as "anti-torque pedals". The anti-torque pedals are located in the same position as the rudder pedals in an airplane, and serve a similar purpose, namely to control the direction in which the nose of the aircraft is pointed.

Pendular action. The lateral or longitudinal oscillation of the fuselage due to its suspension from the rotor system.

Pitch angle. The angle between the chord line of the rotor blade and the reference plane of the main rotor hub or the rotor plane of rotation.

Pressure altitude. The height above the standard pressure level of 29.92 "Hg. It is obtained by setting 29.92 in the barometric pressure window and reading the altimeter.

Profile drag. Drag incurred from frictional or parasitic resistance of the blades passing through the air. It does not change significantly with the angle of attack of the airfoil section, but it increases moderately as airspeed increases.

R

Resultant relative wind. Airflow from rotation that is modified by induced flow.

Retreating blade. Any blade, located in a semicircular part of the rotor disk, in which the blade direction is opposite to the direction of flight.

Retreating blade stall. A stall that begins at or near the tip of a blade in a helicopter because of the high angles of attack required to compensate for dissymmetry of lift.

Rigid rotor. A rotor system permitting blades to feather, but not flap or hunt.

Rotational velocity. The component of relative wind produced by the rotation of the rotor blades.

Rotor. A complete system of rotating airfoils creating lift for a helicopter.

Rotor brake. A device used to stop the rotor blades during shutdown.

Rotor disk area. See disk area.

Rotor force. The force produced by the rotor, comprised of rotor lift and rotor drag.

Rotor tachometer. A device used for indicating the main rotor's revolutions per minute.

S

Semirigid rotor. A rotor system in which the blades are fixed to the hub, but are free to flap and feather.

Settling with power. See vortex ring state.

Shaft turbine. A turbine engine used to drive an output shaft, commonly used in helicopters.

Skid. A flight condition in which the rate of turn is too great for the angle of bank.

Skid shoes. Plates attached to the bottom of skid landing gear, protecting the skid.

Slip. A flight condition in which the rate of turn is too slow for the angle of bank.

Solidity ratio. The ratio of the total rotor blade area to total rotor disk area.

Span. The dimension of a rotor blade or airfoil from root to tip.

Split needles. A term used to describe the position of the two needles on the engine/rotor tachometer when the two needles are not superimposed.

Sprag clutch. Also known as the over-run clutch, a one way clutching device that allows the rotor blades to continue to turn, automatically disengaging the engine in the event of an engine failure.

Standard atmosphere. A hypothetical atmosphere based on averages in which the surface temperature is 59 °F (15 °C), the surface pressure is 29.92 "Hg (1013.2 Mb) at sea level, and the temperature lapse rate is approximately 3.5 °F (2 °C) per 1,000 feet.

Static stop. A device used to limit the blade flap, or rotor flap, at low rpm or when the rotor is stopped.

Steady-state flight. The type of flight experienced when a helicopter is in straight-and-level, unaccelerated flight, and all forces are in balance.

Swashplate. A helicopter swashplate is a pair of plates, one rotating and one fixed, that are centered on the main rotor shaft. The rotating plate is linked to the rotor head, and the fixed plate is linked to the operator controls. Displacement of the alignment of the fixed plate is transferred to the rotating plate, where it becomes reciprocal motion of the rotor blade linkages. This type of pitch control, known as cyclic pitch, allows the helicopter rotor to provide selective lift in any direction.

Symmetrical airfoil. An airfoil having the same shape on the top and bottom.

T

Tail rotor. A rotor turning in a plane perpendicular to that of the main rotor and parallel to the longitudinal axis of the fuselage. It is used to control the torque of the main rotor and to provide movement about the yaw axis of the helicopter.

Teetering hinge. A hinge that permits the rotor blades of a semi-rigid rotor system to flap as a unit.

Thrust. The force developed by the rotor blades acting parallel to the relative wind and opposing the forces of drag and weight.

Tip-path plane. The imaginary circular plane outlined by the rotor blade tips as they make a cycle of rotation.

Tip speed. The rotational speed of the rotor at the blades tips.

Torque. In helicopters with a single, main rotor system, the tendency of the helicopter to turn in the opposite direction of the main rotor rotation.

Trailing edge. The rearmost edge of an airfoil.

Translating tendency. The tendency of the single-rotor helicopter to move laterally during hovering flight. Also called tail rotor drift.

Translational lift. The additional lift obtained when entering forward flight, due to the increased efficiency of the rotor system.

Transverse-flow effect. The condition of increased drag and decreased lift in the aft portion of the rotor disk caused by the air having a greater induced velocity and angle in the aft portion of the disk.

True altitude. The actual height of an object above mean sea level.

Turboshaft engine. A turbine engine transmitting power through a shaft as would be found in a turbine helicopter.

Twist grip. The power control on the end of the collective control.

U

Underslung. A rotor hub that rotates below the top of the mast, as on semirigid rotor systems.

Unloaded rotor. The state of a rotor when rotor force has been removed, or when the rotor is operating under a low or negative G condition.

Useful load. The difference between the gross weight and the basic empty weight. It includes the flight crew, usable fuel, drainable oil, if applicable, and payload.

V

Variation. The angular difference between true north and magnetic north; indicated on charts by isogonic lines.

Vertical vibration. A vibration in which the movement is up and down, or vertical, as in an out-of-track condition.

Vortex ring state. A transient condition of downward flight (descending through air after just previously being accelerated downward by the rotor) during which an appreciable portion of the main rotor system is being forced to operate at angles of attack above maximum. Blade stall starts near the hub and progresses outward as the rate of descent increases.

W

Weight. One of the four main forces acting on a helicopter. Equivalent to the actual weight of the helicopter. It acts downward toward the center of the earth.

Y

Yaw. The movement of a helicopter about its vertical axis.

ACKNOWLEDGMENTS

I wish to thank a few people who helped my flying career, whether they realise it or not, our fun conversations or the serious chats we had and the discussions around flying, made this book possible.

As I worked through the list of everyone who has influenced my aviation career, it is incredible to see the number of people I will always be grateful to. Thank you.

Neville Swan (first gliding instructor)
Craig McNeal (first power flying instructor)
Aaron Shipman
Aaron 'AJ' Jeffery
Aaron Pearce
Aaron Marshall
Adam Eltham
Aiden Campbell
Alan Beck QSM
Alistair Blake
Amiria Wallis
Anastasios Raptis
Andrew Gormlie
Andrew Hope
Andrew Lorimer
Andrew Love
Andrew Sunde

Andrew Telfer
Andy Mackay
Andy Stevenson MNZM
Angelo Cruz
Ben Lee
Ben Marcus
Ben Pryor NZGM
Benjamin James
Bevan Dewes
Bill Reid
Bradley Marsh
Brett Emeny
Brett Nicholls
Bruce Lynch
Bryn Lockie
Carlo Santoro
Carlton Campbell
Chantel Strooh
Charles J. Cook
Chris Barry
Chris Bromley
Chris Pond
Chris Satler
Chris Sperou OAM
Christina Harvey
Christoph Berthoud
Conor Neill
Cosmo Mead
Craig Piner
Craig Rook
Craig Speck
Craig Steel
Craig Walecki
Damien Campbell

Daniel Campbell
Darren Crabb
Daryl Gillett
Dave Blackwell
David Brown
Dave Campbell
Dave Cogan
Dave Hayman
Dave Rouse
David Lowy AM
David Morgan
David Saunders
David Wilkinson
Dennis Eckhoff
Derry Belcher
Desmond Barry
Don Lockie
Donovan Burns
Doug Batten
Doug Brown
Doug Burrell
Dwight Weston
Enya Mae McPherson
Eric Morgan
Eva Keim
Flo Smith
Frank Parker
Gareth Wheeler
Gavin Conroy
Gavin Trethewey
Gavin Weir
Gene De Marco
Geoff Cooper
George Oldfield JP

Giovanni Nustrini
Graeme 'Spud' Spurdle
Graham Lake
Graham Nevill
Graham Orphan
Grant Armishaw
Grant 'Muddy' Murdoch
Greg Quinn
Guy Bourke
Harvey Lockie
Hayden Leech
HH Prince Faisal bin Abdulla bin Mohammed Saud
Ian Lilley
Ian 'Iggy' Wood
Imogen Ling
James Aldridge
Jamie Wagner
Jason Alexander
Jason Haggitt DSD
Jay McIntyre
Jed Melling
Jill McCaw
Jim Rankin DSD
Jock MacLachlan
Joe Oldfield
John Duxfield ARCOM
John Gemmell
John Lamont
John Martin
John McCaw
Jonathan Bowen
Joseph D'Ath
Josh Camp
Juan Ferandoes

Jurgis Kairys
Karl Stol
Keith McKenzie QSM
Keith Skilling
Keith Stephens
Kenny Love
Kermit Weeks
Kevin Langley
Kevin Vile
Kirsty Coleman
Kishan Bhashyam
Kris Vette
Lawrence Acket
Liberio Riosa
Lionel Page
Liz King (Mother Goose)
Lloyd Galloway
Loïc Ifrah
Louisa 'Choppy' Patterson
Malcolm Clement
Martin Schulze
Mark Helliwell
Mark Lowndes
Mary Patterson
Matt Hall
Matt Ledger
Maurizio Folini
Melissa Andrzejewski (nee Pemberton)
Michael Bach
Michael Jeffs
Mike Clark
Mike Foster
Mike Harvey
Mike Jorgenson

Mike Read
Mike Slack
Nando Parrado
Nathan Graves
Nick Cree
Nick Tarascio
Nigel Cooper
Nigel Lamb
Nina Hayman
Paul Andronicou
Paul 'Huggy' Hughan
Paul 'Simmo' Simmons AM CSM
Pete Meadows
Pete Pring Shambler
Peter Harper
Peter Jefferies
Peter Thorpe
Phil Freeman
Phill Hooker
Pip Borrman
Ray Burns
Ray Richards
Reuben Muir
Rex Pemberton
Richard Button
Richard Hectors
Richard Hood
Rev. Dr Richard Waugh QSM
Richie McCaw ONZ
Rick Watson
Rob Fox
Rob Fry
Rob Mackley
Rob Neil

Rob Owens
Rob Weavers
Robert Burns
Roy Crane
Roy Cunningham
Ruan Heynike
Ruth Nisbet
Ryan Brooks
Ryan Francis
Sam Elimelech
Scott 'Macka' McKenzie
Sean Perrett
Shaun Clark
Shaun Roseveare
Simon J Gault
Simon Lockie
Simon Mundell
Simone Moro
SQNLDR Les Munro CNZM DSO QSO DFC JP
Steve Ahrens
Steve Wallace
Stephen Boyce
Stephen Death
Steve Gibson
Steve Newland
Steve Jurd
Steven Perreau
Stu Wards
Tasos Raptis
Tee Jay Sullivan
Tim Marshall
Sir Tim Wallis
Todd O'Hara
Tracy Dixon

Wayne Fowler
Wayne Ormrod
Wayne Thompson
Vaughan Davis
Yoshihide 'Yoshi' Muroya

ABOUT THE AUTHOR

With a passion for aviation passed on from his father who worked in the National Airways Corporation (NAC) office in Auckland, New Zealand, and as a member of the Air Training Corps as a teenager, Fletcher knew all too well the risks associated with flying.

Fletcher often heard about the NAC DC3 Kaimai Ranges crash, as his father's NAC colleagues died in the accident. And as a teenager, the instructor who taught Fletcher how to fly a glider, died some months later in a glider crash. These two instances kick started Fletcher's drive to ensure the safety of those in the aviation industry.

Over his flying carrier, and during his adventures filming extreme aviators around the world, Fletcher tried to understand how pilots ended up in unrecoverable situations. Especially after losing friends to air accidents.

With twenty years of experience working with global entrepreneurs through EO (Entrepreneurs Organisation), training entrepreneurs to experience share between each other and to learn from mistakes, it was a natural transition to help other aviators learn from the lessons of pilots and air crew.

Fletcher selected and compiled these stories to help everyone in the aviation industry learn from others, thus ensuring current and future pilots will be safer in the skies.

www.fletchermckenzie.com

Printed in Great Britain
by Amazon